Endorsements

I believe this is the most important study of hunting since the *Meditations on Hunting* writings by José Ortega y Gasset.

Robin Hurt
Conservationist, Hunter

All true hunters "feel" the truth, but few are able to "articulate" that truth. Now, thankfully, we have *On Hunting* to be our champion of the wild!

Jim Shockey
Naturalist, Outfitter, TV Producer and Host

I very much enjoyed this book. It will have a great reception, as it is about all of us (humans), not just hunters.

Ian Crookston
Engineer, Nonhunter

In the end, a man's worth is not judged by what he knows but by what he has done. Lt. Col. Dave Grossman has done way more for the good of others than he will ever know. His books, all of them, are required reading in my family. His latest book, *On Hunting*, has now joined that required reading list.

Ernest Emerson
"Father of the Modern Fighting Knife," Author of *Bad Guy with a Gun*, Lifelong Hunter
www.EmersonKnives.com

I loved this book! A very intellectual and historical study of why we hunt. Understanding the *why* naturally results in improving skills for tactical situations and survival, not to mention making us better hunters!

Carl Chinn
President of Faith Based Security Network, Lifelong Hunter

ON HUNTING

A DEFINITIVE STUDY OF
THE MIND, BODY, AND
ECOLOGY OF THE HUNTER
IN THE MODERN WORLD

Lt. Col. Dave Grossman, Linda K. Miller, Capt. Keith A. Cunningham

BroadStreet
PUBLISHING

BroadStreet Publishing® Group, LLC
Savage, Minnesota, USA
BroadStreetPublishing.com

On Hunting: A Definitive Study of the Mind, Body, and Ecology of the Hunter in the Modern World
Copyright © 2022 Lt. Col. Dave Grossman, Linda K. Miller, and Capt. Keith A. Cunningham

9781424564927 (softcover)
9781424564934 (ebook)

Scripture quotations marked CEB are taken from the Common English Bible, copyright © 2011 by Common English Bible.

Stock or custom editions of BroadStreet Publishing titles may be purchased in bulk for educational, business, ministry, fundraising, or sales promotional use. For information, please email orders@broadstreetpublishing.com.

With thanks to literary agent Richard Curtis.

Cover image of rifle courtesy of Linda K. Miller.

Cover and interior by Garborg Design Works | garborgdesign.com

Printed in China

23 24 25 26 27 5 4 3 2 1

Dedication

To all hunters, today and past,
who have informed our hearts and minds
and to those who follow the hunter's path after us

A Brief Note on Gender

We often think that hunting narratives lend themselves to male examples and the male gender. There are significant exceptions to this rule. Some magnificent female hunters are presented in this book, but in general, the authors will refer to hunters using the male gender.

This is in no way intended to exclude females from the history or the future of hunting.

Indeed, many mythological deities were "goddesses of the hunt," and recent research indicates that early big game hunters in the Americas were both male and female. This challenges the long-standing myth of "man the hunter" or "woman the hunter." It is truly "human the hunter."

Buffalo Jump

An aboriginal guide was showing a group of tourists around [the exhibit at] Alberta's Head-Smashed-In Buffalo-Jump...The guide graphically described how in ancient times the buffalo would be driven over the edge of a fifteen meter precipice, to land in a gory heap at the base of the cliff. A diorama showed men and women clambering over the bodies to club and spear those still living.

When one tourist expressed shock at the bloody nature of the enterprise, the guide responded simply but with conviction, "We were hunters!" connecting her own generation with those of the past. She then amended her statement with equal conviction, adding, "Humans were hunters!" thus expanding complicity in the act of carnage to the whole of humanity [including the tourists].

—Richard B. Lee and Richard Daly,
The Cambridge Encyclopedia of Hunters and Gatherers

Table of Contents

Foreword

Lt. Col. Dave Grossman is a renowned author, and I'm the happy owner of many of his wonderful books. His work focuses on human performance factors under stress. They enlighten us on the human condition before, during, and after a personal defense situation. I've always had great respect for his work, especially his audiobooks and live seminars. So I was excited to hear that Dave's new book, with authors Linda K. Miller and Capt. Keith A. Cunningham, was about hunting. My husband, Ken Ortega, and I have had some of our best times together hunting.

I believe that hunting is a great common ground for all of us. Man, woman, or child, we are equals when using the same caliber! I have had the honor of sharing my defense training and my hunting knowledge with both new and experienced hunters. I have trained men, women, and children from all over the world in the great joy of hunting their own food. Many of these opportunities came through my roles as pro staff (Bass Pro Shops/Cabela's) and as a hunter education instructor (state of Nevada).

All that to say, my ability to provide for my family through hunting is a joy. It allows me to bring organic meat to the table. And I love to help others discover and experience hunting as well.

Thus, this book resonates with me. It helps people understand why we need wildlife conservation. It helps even nonhunters understand the virtues of the hunt. It helps the timid souls who buy their meat in neat little packages at the grocery store understand the field-to-table process in the most fundamental way.

I look at hunting as a rite of passage. Some important parts of the process are training to use the tools and understanding that an ethical, humane harvest helps us to participate in wildlife management for a healthier future. In some cases, a managed harvest can even save animals from extinction.

This book has a very easy way of putting those things into perspective and presenting the truth about hunting. When we hunt, we have a moral and ethical responsibility to do our best. We must select the right animal. We must be able to place the shot to ensure a quick, humane death. And we must honor them without haste or waste while thanking God for providing us this bounty.

The truth of why we hunt is also described in this book. Ultimately, *On Hunting* provides the foundation for an honest conversation about hunting in the modern world.

I hope that you will enjoy *On Hunting* as much as I did and apply its principles to your own life and worldview. Surely, all hunters and conservationists will appreciate and enjoy it. *On Hunting* will impact us for generations to come.

Maggie Mordaunt
"CCW Maggie"
Female Hunter
Pro Staff for Bass Pro Shops/Cabela's
Nevada Hunter Education Instructor
Co-Owner of Homeland Personal Protection, LLC

Introduction

Bruce Siddle's research on predation shows that, according to British records, in India from 1900 to 1910, over one hundred thousand people were killed by tigers.[1] Those are just the ones that were recorded in a single decade. Imagine what it must have been like a century before that. Or a thousand years or even ten thousand years before that, when our species must have lived a life of constant predation.

To understand mankind, we must understand that throughout our history we have been in the middle of the food chain. We have the gripping fangs and the forward-set eyes of a predator, and in our brain, we have the neural pathways of a team of cooperative, goal-oriented, and (even, perhaps) self-sacrificing predators. We also have the chisel teeth of a rabbit, and we have the neural pathway of a "blow the ballast" (mess yourself) and run like hell rodent. Finally, we have the grinding molars of a grass eater and the neural pathway that says, "I don't have to outrun the predator; I just have to outrun the slowest one in the herd."

> You're the predator right up until you're prey.
>
> —James S. A. Corey, *Abaddon's Gate*

We have these various survival responses built into our bodies, our minds, and our genetics. But which is the most fun? It is

11

no fun to be prey! We are never genuinely happy when we are prey. It can be argued that we are at our happiest and healthiest when we are the hunter, not the hunted.

Even when we made the switch from hunter-gatherers to agriculture, we continued to be hunters. To this very day we see this in rural parts of our society, where modern farmers and ranchers also, and almost universally, consider hunting as an important part of their lives and their livelihood.

The history of humanity is a long struggle to survive and to claw our way to the top of the food chain. We have within ourselves the capacity to be prey or predator. Thus, you can make a good argument that one of the great pathologies of modern times is an abandonment of our predator roots. A healthy mankind embraces all aspects of our heritage and our genetics.

The basic premise of this book is that we are what we were when we were formed.[2] And we (*Homo sapiens*) emerged when we were all hunters. No matter how we socialize ourselves, "under the covers," we are all wired the same way. We were all hunters then, and we are still.

In this book, we explore

- the "Deep Roots" of hunting, how it is woven into our very core;
- "The Hunter's Oath" and the ethics of hunting;
- "Kudu Eyes," the underlying skills of the hunter;
- "It's What I Do," hunting skills applied to police, military, and self-defense; and
- "It's Who I Am," the challenges, traditions, and reverence of the modern hunter.

Overall, this book is a celebration of life, of the hunter's life. It is an acknowledgment of our place in the middle of the food chain as both predator and prey. And especially, it sheds light on how these two roles (striving to get food and striving not to become food) are inextricably linked to procreation. Many people feel guilty that they have a sense of euphoria when they are successful in the hunt, but the authors of this book believe this is normal and natural. It is our brain's way of responding when we are confirming our life force and improving the odds that our species will live and multiply.

> All animal brains are designed to create flashes of pleasure when the animal does something important for its survival.
>
> —Jonathan Haidt, *The Righteous Mind*

Artemis

A bridge from our earliest hunting roots to our modern souls is the transfer of ancient hunting stories to the mythologies of our early tribes and civilizations. In Roman mythology, it's Diana. In Greek mythology, it's Artemis, a profoundly important member of the Ancient Greek pantheon. She was a very popular goddess in her day, widely admired and worshiped.

She is probably best known now as the goddess of the hunt. What more important concern can there be for primitive humans than the hunt, wild animals, and the wilderness? This was the realm of Artemis. Not only was her domain the wilderness, but she was also considered to be wild, particularly in the sense that she was fiercely independent and comfortable with being alone. She refused to conform to convention or tradition. She was the personification of the wilderness: self-sufficient, living by her own terms, both goddess and woman.

Her portfolio also included the protection of young girls, virginity, and childbirth. Perhaps, in the transition from innocent virginity to childbirth, ancient humans saw the essential life that flowed from that natural process. She was a goddess of great compassion, but she punished mercilessly those who needlessly harmed creatures under her protection.

Like the hunt leader that would evolve into the war chief and the civic leader, she was comfortable both in solitude and in leadership. She rejoiced in the hunt while simultaneously being the guardian of nature and wild animals. (Another powerful parallel to modern hunters!)

Artemis was the embodiment of the skillful, proficient, free-spirited, self-reliant woman.
She was then—and is today—a role model for the capable, strong-willed, and independent female.

It is no accident that Diana/Artemis ruled the wild animals, the hunt, and procreation. These are the basis of life: birth, predator, and prey.

> The circle of life…that's what hunting's all about.
>
> —Jim Shockey (writer, photographer, guide, and outfitter)

Campfire Story: Her First Deer
Another of the bridges from our ancient hunting roots to our modern souls is in the telling and retelling of hunting stories around the campfire. To genuinely enjoy these stories, the reader needs to slow down, sit back, and relax.

It was movement that caught her eye. Her grandfather had told her it would most likely be movement that she would see first—at least that was the way it worked when the second doe came into the clearing. It had stood on the edge of the tree line while lowering and lifting its head, flicking its ears, and twitching its tail. At this time of late autumn, there was very little feed that could be described as lush, so this grassy little clearing was attractive to the deer.

Earlier, the first doe had just appeared. The girl had scanned the clearing, as was becoming her routine, and she had seen nothing. She had glanced out another window in her blind to locate the red squirrel that was scolding her for being in its territory. Its bushy tail vibrated in the cool morning air as it chattered its disapproval. She smiled at its antics, and when she turned back to do another scan, the doe was just standing there in the middle of the clearing. How was it possible for it to get that far without being seen?

She had gasped softly as the small shot of adrenaline hit her bloodstream but calmed quickly, realizing that this deer was not what she was after. Her grandfather had told her this would happen, and she took several deep breaths to settle down. He had called it "combat breathing." It had worked for him when he was in combat, and it worked for her now.

Suddenly, there he was! This deer had antlers, big antlers, and could be the one. *Breathe…breathe*, she told herself. The buck hadn't seen her and was far more interested in the doe.

Now the buck moved out into the clearing. He was magnificent. The girl now experienced a shot of adrenaline like never before. This was it! This was the buck! She had prepared herself for just this moment, and now she was shaking with excitement. *Breathe…Breathe!* she screamed in her mind. She could feel herself calming as she put her sight reticle on the point of aim.

She could see the sight reticle was steadier now. When the big deer stopped to test the wind, she was ready to fire her shot.

It was a good shot—right on the point of aim. She and her grandfather had practiced making shots just like this. She had fired many such shots on paper and steel targets before this one all-important shot.

The buck immediately did a mule kick and ran toward the nearest tree line. Then the deer slowed, became unsteady, and finally collapsed. She could see him thrashing, but she could tell it was over. She waited a bit longer before going over to him.

As she approached, she kept her eyes on the one antler she could see, with her rifle in a "ready for anything" position. She circled slightly to approach the deer from its back and could smell the odor of a buck deer in full rut. She nudged him with her foot, and there were no signs of life. She then moved around to look into his eyes and saw the stare of death.

She knelt by his head, patted him lightly and said, barely more than a whisper, "Thank you…thank you."

She sat quietly thinking about her first deer until she felt her cell phone vibrating. It was a text from her grandfather; he had heard the shot and was wondering if it was time for him to come and help.

"Yes, Grandpa," she replied. "And Grandpa…I love you."

> Hunting is what we were designed to do. I am not saying it is a good thing or a bad thing; it is just the way we developed or were created (depending upon your perspective). We can no more deny it than a bird can deny its wings. We can no more ignore it than we can ignore our canine teeth or our flesh-ripping incisors. It is who we are. It is what we are, and it is foolish to try to pretend otherwise.
>
> — Lt. Col. Dave Grossman

Introduction—Summary

So, are you predator or prey? The truth is that we have within us aspects of both. But we have clawed our way to the top of the food chain because *we do not like being prey*! We are *wired* to be happy and healthy when we hunt. As we will see in the next section, the hunter is who we have been and who we desire to be.

Section I

Deep Roots

If we go back far enough in time, to our roots, we see that hunting is what allowed us to survive and become who we are biologically, neurologically, and physiologically. We were hunters from the beginning. Our ancestors fed themselves by hunting. And it was the successful hunter who got to procreate. While we may not consciously remember being hunters, our genes, our bones, and our brains are a record of our past. Even today, we see our ancient form reflected in our mythologies, our rituals, and our traditions. Our past built us. We have deep-seated yearnings to connect with our ancient selves. Most of us strive for that connection, generally without knowing what we are really yearning for.

Chapter One

We Were Hunters Once, and Prey

When we first started to research the history of hunting for this book, we were thinking about the history of civilization. We were thinking about a history that stretched back to the emergence of populations in Mesopotamia about five thousand years before the current era. We soon realized that this was the history of modern, literate civilization. It was a history that didn't address the many years our ancestors lived and explored and hunted before they ever gathered in settlements and started writing things down.

So, in this first chapter of the book, we explore our ancient history. It is the history of humanity from before we began our modern lives.

We were all hunters then. We needed to be predators to feed ourselves. Our ancestors, bless them, found ways to dominate the food chain. They could then eat the quantities of protein that fed our magnificent brains.

A Short History of Man
We are not embracing an evolutionary standpoint. The authors deeply respect all perspectives on this subject. But a brief look at

hunting from an anthropological outlook can give us an astounding insight for those who would consider the matter from an evolutionary perspective:

Anthropologists date the first appearance of early humans at about 2.5 million years ago.

They date the emergence of modern man at about two hundred thousand years ago.

They say that twelve thousand years ago, every human was still a hunter-gatherer.

Gradually, hunter-gatherers turned to agriculture, and by 3500 BC (early Bronze Age), most humans were farmers. (Many still relied, at least partially, on hunting.) Only in very recent times has farming declined, with more people now living a commercial-industrial urban life. Another way of looking at this enormous time span is to think in terms of a single day. If early hominids appeared twenty-four hours ago, modern man appeared less than two hours ago. Less than six minutes ago, we were all still hunter-gatherers. About three minutes ago, most of us were farmers, and less than thirty seconds ago, most humans became commercial-industrial urban dwellers. So, we are relatively new at being urbanites, and our past as hunters is what truly grounds us.

Thus, whether we evolved that way or were created that way (or any of the worthy perspectives in between), we can recognize that our brains are still dialed in to the hunt. None of the diverse standpoints on this subject can deny that we are, in many ways, a product of the "survival of the fittest." Nature's merciless realm of the "law of the jungle" has had an enormous impact on who we are and who we must be in order to survive. We are designed to hunt, and our mind is wired for the hunt. Our vision is tuned for being both predators and prey. Our hormones are geared to finding food and not being food. And while they may seem out of place in our current lifestyle, many of our strengths are designed for our success as hunter-gatherers.

Inside each of us are the ancient brain and the hormones of the hunter. Inside each of us is our heritage. And to feel whole, we need to see the hunter within, to acknowledge him, and to celebrate him.

> In order to subsist, this early man had to dedicate himself wholly to hunting. Hunting was, then, the first occupation, man's first work and craft.
>
> —José Ortega y Gasset, *Meditations on Hunting*

From Hunter to Human

So, who were we for all those years? Anthropologists frequently make new discoveries about our ancient roots and our hunting methods. First, we were nomadic hunter-gatherers. We moved around to follow game animals and to take advantage of wild food harvests. It appears that we started with some bare-hands hunting (using ambush and persistence tactics).[3] Later, we progressed with tools (clubs and spears). Like modern-day hunter-gatherers, we likely ate a lot of small game, but when we had a large kill, our entire village-family would move to the kill site (at least for a time). The present-day !Kung Bushmen still do this.

> It is more than clear that these men...the most ancient that we have glimpsed, did not invent hunting but rather received it from their...past.
>
> —José Ortega y Gasset, *Meditations on Hunting*

Archeological evidence shows that we've been hunting for meat for a very long time. In fact, most of the history of man is a story of honing our hunting skills. As Robin McKie, science editor at *The Guardian*, wrote:

Evidence from ancient butchery sites in Tanzania shows early man was capable of ambushing herds... [and using] complex hunting techniques to ambush and kill antelope, gazelle, wildebeest and other large animals at least two million years ago. This discovery—made by anthropologist Professor Henry Bunn of Wisconsin University—pushes back the definitive date for the beginning of systematic human hunting by hundreds of thousands of years.

McKie continued:

Bunn believes these early humans probably sat in trees and waited until herds of antelopes or gazelles passed below, then speared them at point-blank range. This skill, developed far earlier than suspected, was to have profound implications.[4]

And as *The Cambridge Encyclopedia of Hunters and Gatherers* says: "Hunting and gathering was humanity's first and most successful adaptation, occupying at least 90 percent of human history. Until 12,000 years ago, all humans lived this way."[5]

Man the predator has a long and proud history. Plentiful game made us who we are today: large-brained hunters with a taste for meat.

> Neanderthals had many healed injuries to their skeletons resembling those seen in rodeo cowboys, suggesting some very rough activities, perhaps killing large animals with hand-held weapons instead of projectiles like hurled spears or arrows.
>
> —Peter J. Richerson, University of California, Environmental Science and Policy

The Hunting Hypothesis

As far back as we can imagine, our ancient ancestors must have eaten everything. They ate nuts and seeds, grubs and locusts, sprouts and tubers, as well as mice and antelope. Hunting increased the proportion of meat in their diets. And meat, especially cooked meat, powered the brains that make us "*sapiens*," or wise.

The ancient barbecue could well have played a role in developing our social interactions. Fire kept us warm and held predators at bay. And, while it's easy to cook chunks of a large animal on a campfire, cooking smaller critters and vegetables may have made pots necessary. Although our ancient ancestors were no doubt nomadic, both fire and pots would tend to keep them in one place a little longer. Relocation would now involve moving "stuff," or at the very least, protecting the coals they would carry to the next campsite.

The campfire could have shaped social behaviors such as tribal bonding, family structures, and mating. Along with the emergence of language, these would have accelerated our ancestors' development. Importantly, these social behaviors would also have improved our hunting success.

So, did hunting improve our social behaviors? Or did social behaviors improve our hunting? It is not either-or…it is "yes" to both. Hunting reinforced our skills in social behavior, and social behavior reinforced our skills in hunting. Family structure and tribal bonding developed in support of our hunter-gatherer activities. There is certainly strong evidence that our ancestors mated successfully. This likely involved extended family groups. Extended families would help tend the babies *and* allow for successful hunts.

Language, or voiced communication, is not the exclusive domain of modern man. While we know that many animals voice "words" that they use systematically, much of their (and our) communication would likely have been non-verbal, as is common in the animal kingdom. Early words were probably not much more than monosyllabic sounds. Mime may have been an important part of communications. Picture a scout-hunter describing the size, direction, distance, and location of a herd of antelope. Perhaps "written language" started with drawing a map in the sand, with symbols for the type of animal and the size of the herd. The cave paintings that have survived the ravages of weather and time tell us that stone-age man's primary interest was animals. But many of these paintings also include drawings of symbols, such as waves, squares, open angles, crosses, hashes, and similar signs. What do they mean?

Anthropologist Genevieve von Petzinger is renowned for her work in recording and interpreting cave art. An article on cave paintings in *The Guardian* reported on the significance of her analysis:

> "What we found was quite remarkable," says von Petzinger. "There is definite patterning in the way these signs were used." In other words, she and [her colleague] Nowell have shown that these markings are no mere abstract scribbles but appear to be a code that was painted on to rock.[6]

And if that weren't amazing enough, "Many of the swirls, crosses, circles, open angles and crosshatches seen in France are also found in...earlier works from Africa."[7]

There's also likely a link between the development of religion and hunting. The notion of "religion" is probably a very modern concept. Our hunting ancestors might well have called it "science" or a "description of how the world works." As David Petersen says, "For pre-agricultural foraging peoples—our 'savage' human forebears—sacred and secular were inseparable."[8]

There is no specific evidence about what our pre-literate ancestors thought. We can only speculate. But it's quite possible they believed that animals possessed a spiritual essence. Anyone who has hunted knows that there is a qualitative difference between the live animal and the carcass. It's not just that the animal stops moving; it's also that the animal's eyes look different. And then, when the animal is being field dressed, there is a palpable change...the sound of escaping air and fluid, the heat and sometimes steam that rises from the guts, the somewhat sweet smell of the blood. Certainly, it's easy to understand why our ancestors might have believed that a life-spirit was leaving the animal. And it is easy to understand that modern hunter-gathers often practice what we call "animism," the belief that nonhuman entities (animals, plants, and natural phenomena) possess a spirit or a spiritual essence.

> There is a passion for hunting,
> something deeply implanted in the human breast.
>
> —Charles Dickens, *Oliver Twist*

It's easy to see that the cave paintings were, in part, an expression of that belief. Animals were a central feature of their lives, and the drawings could have been used as a "storyboard" describing past hunts and teaching new hunters. The cave art might have been a "vision board" that focused their efforts for future hunts. Those of us who have spent time hunting know that it's easy to start dreaming about the animals long before we go out. The modern hunter usually pores over pictures from trail cameras. Tales of great hunting moments of the past fill the hunt camp. It is very much like an echo of the cave drawings of our ancestors.

By the time there were written records (about five thousand years ago), the great ancient civilizations (Egyptian, Indian, Chinese, and many more) had started to arise. Each of these cultures formalized their pantheon of deities. Most included a god or goddess of the hunt, and these were often identified with a specific animal. Diana the Huntress usually stands with a deer or hunting dogs. Artemis favors a stag. Cernunnos (Celtic god of the hunt) rules with a ram-headed serpent and a stag.

Many of the hunting deities of mythology were female. Their companion animal was often a stag. Ironically, paleoanthropologists propose that the role of women during our hunter-gatherer days was primarily as a gatherer. Modern scientists suggest that pregnancy and child-rearing were the women's dominant roles. They suggest women would have hunted small animals (insects, grubs, snakes, small mammals) that they could catch easily and close to home (with children in tow). Researchers caution us that what we see in some modern hunter-gatherer cultures (where the women typically hunt small animals) may not reflect what our ancestors did.

In fact, more recent research (looking at burial records throughout North and South America) has shown that ancient big-game hunters were of both genders. They estimate that 30–50 percent were female.[9] Prior to this finding, scientists thought that the women typically hunted low-risk, high-average-yield small animals (usually used to feed the immediate family), and the men hunted high-risk, high-single-yield big game (usually shared among a larger community). Further, the scientists suggested that the largess of the successful hunter (returning to the community with enough meat for everyone) would be socially important. It would earn him reciprocity from the other hunters (and give him insurance against less successful hunts). And it would give him status (yielding more access to female favors).

The "boom and bust" cycle of big game hunting may also have necessitated the preservation of meat by drying or smoking (or in cool climates by fermenting and in colder climates by freezing). And again, modern paleoanthropologists suggest that any care of meat products was the domain of the female, but this is largely conjecture.

And no matter how modern eyes interpret the past, humans were hunters, and both genders were hunters at least to some degree.

> How, given the canine teeth and close-set eyes that declare the human animal to be a predator, had we come up with the notion that oat bran is more natural to eat than chicken?
>
> —Valerie Martin, *The Great Divorce*

And Prey?

Humans were (and are) in the middle of the food chain or, as some scholars put it, the "food web." We are predators, yes...and we are also prey. As prey, we were then (and we are still) highly vulnerable. We can't run as fast as other prey animals. We don't have fearsome fangs and claws. We don't have thick hides. We die easily and quickly. Altogether, we are vulnerable prey that's easy to eat. With no thick fur, the first bite is all food. Humans are really "soft targets."

We often forget that we were once prey and still are prey. Every year there are many stories about animals attacking humans. In many areas of Africa, this is vividly true. Lions, leopards, hippopotamuses, and elephants (and less often, chimpanzees) regularly attack people. Crocodiles also regard us as prey, no matter where they find us. From south to north, bears (especially polar bears and grizzlies) and all the big cats think of us as prey.

One paper presented at the Gordon Research Conferences (an international forum for frontier research) summed it up:

Predator-prey interactions have shaped all life on earth, and it is this underlying commonality that helps explain the development of so many parallel, but as yet, independent research paths…Though we tend to think of predator-prey interactions as pertaining to other species more than humans, there is increasing recognition that many aspects of the human condition have been shaped by our…history as both predators and prey.[10]

> We are wholly dependent on the organisms who pass life along to us…We are animals participating fully in the ecological network…We hold membership in the environment no less than any other species.
>
> —Richard K. Nelson, "Finding Common Ground" in *A Hunter's Heart*

Everyone who has hunted has also interacted with their prey. Every creature has its "comfort zone." The partridge flushes at about ten yards of our approach. The black bear avoids us at about fifty yards. The white-tailed doe puts her tail up at about one hundred yards. If you get too close, you will evoke their "prey response."

We humans have this "prey response" too. One of the basic reactions that we have (and have had throughout time) is the "fight or flight" stress response. Our brain chemistry helps us respond to stress by mobilizing all the forces we might need either to fight or flee from a predator. In the moment of aggression, chemicals like adrenaline flood into our bloodstream. Our breathing increases. Our blood goes to our large muscles. Our attention intensifies. Our reaction time quickens. Our pain perception decreases. We are ready for the fight! Or we are ready to run away from the threat! What's interesting is that when we hunters are stalking in the field, we are demonstrating classical "predator response." This is one of the most enjoyable aspects of hunting.

As researchers said in an article published by the National Center for Biotechnology Information:

> Thus, the confrontation between predator and prey is an antagonistic dance, involving common actions and competing motives that may be part of humans' mixed emotional inheritance.[11]

The lion and the lamb may indeed lie together in our brains.

> The essence of wildlife is wildness.
> And the heartbeat of wildness is predation.
>
> —David Petersen, "Hunting as Philosophy Professor" in *Hunting—Philosophy for Everyone*

We come by our predator-prey blend honestly. These traits come from our deep past. They are badges of our ancestors' survival in a predator-prey world.

It is easiest to see the physical characteristics that we inherited and sometimes a little harder to identify the mental and emotional ones. Some are still assets, and some may seem outdated. They are all important in understanding our current selves and the journey that got us here.

- Our teeth still include the signs of meat-tearing fangs.
- Pre-verbal alarm calls (screams) stay with us from infancy through our adult lives.
- "Hair standing on end" or "goose bumps" are muscle reactions that make all primates' fur stand up and make them appear larger and more threatening to a potential predator.

- Our eyesight is highly tuned. Notably, it provides the binocular vision that gives us depth perception and the accurate distance assessment we need as predators. Even with our forward-pointing eyes, we also have good peripheral vision, which is the hallmark of the prey animal.

- Animals capture our attention. In tests, researchers found that we are significantly better at noticing changes in pictures of people and animals than we are at noticing changes in inanimate objects, even though inanimate objects are the primary hazards in modern life.[12]

- Aggression is a modern trait that would have been an important aspect of survival for our ancestors, not only to capture their prey but also to threaten those who preyed on them.

- Humans maintain a general level of anxiety that aids our situational awareness, useful for both predators and prey. But in modern life, it often has no outlet and may contribute to many of the modern stress-related diseases.

- Intelligence is a particularly human trait that was probably our best offense and our best defense in the wilderness.

- Human social behavior and cooperation likely helped us achieve mutual protection and security and very likely helped in the hunt.

> Meat supplied a more concentrated package of calories and
> nutrients than weeds and berries. Not being the biggest
> and strongest members of the food chain, however, *Homo
> carnivorous* also required more cunning and wile to bring
> down that mastodon. One theory holds that a bigger brain
> and a longer period of nurturing and apprenticeship had to
> evolve to master the hunt.
>
> —Gary Stix, "Homo Carnivorous" in *Scientific American*

Tool development was a major factor in giving early humans a predatory advantage and may also have helped in driving away animals that threatened their security. As a predator, the human without tools is very much at a disadvantage. He has thin skin, no claws or fangs, weak bite strength, weak gripping strength, relatively light bones, and relatively poor overall physical strength compared to other apex predators.

However, our social skills, intelligence, and tools eventually gave us the advantage.

> Hunter and hunted are partners, the presence of each honing
> the instincts and senses of the other. The hunter possesses the
> skills to chase; the hunted the skills to evade. Each provides
> the other with an energy carefully balanced. Life, through
> death, provides life. The predator is integral to the prey.
>
> —Ruth Rudner, "The Call of the Climb" in *A Hunter's Heart*

Presumably through selection of the fittest (and de-selection of those who could not defend or flee), we humans developed a superb set of responses to fear (or stress or excitement). Since most of us no longer must outrun a lion, sometimes these reactions may seem to be a little strong for the situation. (You receive a late-night email from your boss, and your heart rate jumps. You don't really need more blood pumping through your veins to read an email, but there it is.)

Without technology, we have little advantage in a fight against animals with fangs and claws. Ancient man lacked much in the way of technology. A club is not much of a defense against a charging predator. Ancient man had only one real advantage: his wits. And, by eating a protein-rich, mostly meat diet, he fueled the highest-functioning brain in all the animal kingdom.

Amnesia

Over time, with the advent of agriculture and then modern urban life, most people forgot their hunting roots. Within a few hundred years, we lost our conscious memory of the previous thousands.

Ironically, modern people often dislike the very things that made them who they are. The diet that fueled our magnificent brains has become unfashionable. Modern people often fear the technology of hunting. They sometimes see the "fight or flight" stress reaction as a liability. Natural aggression seeks a socially acceptable outlet. Many of the skills and traits that stimulated our rapid rise in the animal world, when misdirected or exaggerated by modern life, can lead to human reactions, such as anxiety attacks, hyper-vigilance, and misplaced aggression. Is it possible that not having a productive outlet for these skills can cause pathologies?

The predator-prey world designed our brain chemistry. While we've become more cerebral, our bodies are responding as though we still live in a world that's populated with persistent predatory threats and irresistible prey opportunities. For many people, this feels like having a hammer but no nails. This sense of being the right person at the wrong time produces an unfulfilled yearning…but a yearning for what?

Indeed, humans must have influenced the very evolution of large mammals like deer, elk, moose, buffalo, antelope, and caribou; and so, we ourselves helped to create the qualities we love so deeply in these creatures: speed, grace, agility, elusiveness, strength, and wildness itself.

—Richard K. Nelson, "Finding Common Ground"
in *A Hunter's Heart*

Chapter One—Summary

A brief review of human history demonstrates that modern humanity is just a thin veneer over our vast history as hunter-gatherers. Until very recently, hunting has been to us what it is to the lion pride or the wolf pack.

Hunting is who we were and who we are, and it is deeply embedded in our brains and our culture. Hunting formed us. It is tightly woven into every aspect of that which defines us. Hunting is inextricably embedded in our eating, our thinking, our tool-making, our speech, our social structures, our art, and our beliefs. Hunting defines what it is to be human, and we cannot understand humanity without understanding hunting.

Chapter Two

A Guilty Pleasure

In this chapter, we look for connections between our ancient past and our modern life. We are looking for connections that may help to restore our collective "amnesia." We are hoping to unite present-day humanity with our hunting past. We want to remember how we found a way to kill and eat meat. We need to understand why we wanted to do it over and over again.

Going "back to your roots" is a popular saying. It usually means going back to your place of origin, going back to where you grew up. But it also means going back to where you felt connected, connected to the things that are important to life itself. Going back to your roots also means returning to who you really are, returning to your life source.

Today, there's a wave of interest in "heritage travel." This brings travelers to geographic locations that are closer to their origins. In a *Forbes* article, Tom Marchant writes that heritage travel "can be one of the most re-affirming experiences in life and give travellers a greater sense of history and identity."[13] The idea of an "eco-vacation" is often similar. This is where the traveler spends a week in the jungle or another remote spot. You're away from the digital world without electricity or running water. You're connecting to an earlier time in human history.

And hunting is something that's even closer to our origins than heritage travel or a week without modern luxuries. However, it is increasingly challenging to find a place to hunt. Cities, suburbs, farms, and summer cottages cover much of our land. Often, private owners control most of the bush land. The government manages much of the forest. Even with the difficulty of finding a place to hunt, there's a genuine revival in hunting of all types.

Women are driving this revival. Many people speculate why this is so, and there are likely a variety of reasons. Women may perceive hunting as a "man's world" and as another frontier to explore. But women who try it and like it say there's more to it. As Alanna Mitchell reports in a *Toronto Globe and Mail* article, one young huntress says, "It's uplifting to be part of it, to be self-sufficient."[14] The feeling of self-sufficiency is empowering. With empowerment comes confidence. For some women, hunting generates a primal response. They are electrified by the experience of being one with the "circle of life."

For both men and women, hunting is not just about food. It reinforces their connection to nature. The act of hunting, and especially the kill, is one of the most intimate, primal, respectful acts in which a human can take part. Hunting vividly reinforces the importance of being able to take care of oneself and, even more, being able to provide for the family. As the old saying goes, there's nothing like being able to "bring home the bacon."

For women who hunt, and especially if they hunt big game, it's difficult to articulate their reasons and motivations. They often call the experience indescribable. This speaks to the mystic quality of both the hunt and the kill. It is deeply and mysteriously spiritual. Some call it intimacy. Some say it is the closest they've ever felt to nature (or God) and to what it means to be human. Others say that the world stands still for a time.

It's the same for men as for women. Men usually can't (and often won't) articulate the feelings they have when hunting. It's not machismo; quite the opposite. It's the excitement you feel while fully engaged in hunting. It's a vibrant connection with reality that's almost super-real. Some describe it as a mystical experience. They call it a feeling of unity with the absolute (whether God, nature, or the universe). They say they achieve a spiritual understanding of truth that's beyond the intellect.

And when the game approaches, the excitement can feel almost unbearable. You try to stay calm. You focus on the job at hand. Yet your leg muscles tremble. Your heart pounds in your ears. You have to command yourself to breathe!

When hunters talk to hunters, they know. Each knows whether the other has felt that primal response, one that taps into the core of what it means to be human. The surge of adrenaline provides a thrill that goes from your hat to your boots. It makes the hunter feel fully alive and connected to a primitive reality, a connection that spans all time.

That thrill can be addictive. If the hunter's usual game doesn't provide that adrenaline hit, he may decide to change the game and choose a more dangerous type of animal. He may change the location to an exotic place such as the jungles of Africa. He may change the technology from rifle to bow to spear. Many people lack the budget for dangerous game hunts. So, they watch the TV shows and read the books about exotic hunts. And they have regular fantasies about delivering the final shot to the charging Cape buffalo. Or they imagine themselves saving the day as the roaring lion makes his final leap toward the hunting party, the human prey.

On Killing

New hunters especially must come to terms with the fact that they intend to kill an animal. If they are tightly tied to the idea of the "circle of life," they quickly accept that a successful outcome for them means an unsuccessful one for their prey. Otherwise, they may have to work through feelings of anxiety or guilt until they come to appreciate that some things die for others to live. Eventually they are ready to take full responsibility for sourcing their own nutrition. They see hunting as an alternative to outsourcing their shopping for meat to a food chain. They see that their meat doesn't need to come in plastic wrap from the grocery store. And they will find hunting moves them a step closer to their roots. It moves them a step closer to fulfillment as a human animal.

But it's killing, isn't it?

Then why does it thrill you?

Many people have heard the expression "the thrill of the hunt." They think it means that the process of seeking the game is thrilling…and it is. But the big rush occurs at the moment of climax. As the game animal draws near and enters within range, "buck fever" can overcome the hunter. The brain dumps chemicals into the blood and produces the fight-flight response. The hunter struggles for self-control. *Breathe*, he says to himself, in an effort to calm his body and his mind. Then he shoots a shot or looses an arrow. And a huge rush, a euphoria floods over him.

Why do hunters kill?

José Ortega y Gasset, author of the much-quoted *Meditations on Hunting*, is often recognized as the father of modern hunting ethics. In his *Meditations* he says, "One does not hunt in order to kill; on the contrary, one kills in order to have hunted."[15] We hunt for the thrill of the chase and the ecstatic peace that comes with being out there trying to beat a wild animal at his own game.

When the chance finally comes, there is no doubt; we will kill. But it's not about being a killer.

Asking a hunter why he kills is like asking a fish why it swims or a bird why it flies. It is defining. It is ancestral. It's "in our blood."

"Hoo, boy!"

The chemical cocktail that the brain dumps into our bloodstream during the hunt, and especially during those moments leading up to and following the kill, is the definition of what it means to be *Homo sapiens*. This process of dumping chemicals has reinforced our response to fear, stealth, and excitement, all of which are part and parcel of the hunt…and of being human.

Most of us are familiar with adrenaline. It and related neurochemicals work together to give us an instantaneous response to a stressor. It increases our blood pressure, pushes blood and oxygen to the big muscles, increases our energy, and focuses our attention.

Cortisol is another stress hormone but one that takes a little longer to act (minutes, rather than seconds). It helps regulate body fluids and channels energy away from nonessential body systems (nonessential to the stress response, that is). Like most aspects of the stress response, it's very good when you're in a predator-prey situation and very bad when it becomes a chronic reaction to daily living. In an article at *HuffPost*, Sarah Klein explains:

> When you stew on a problem, the body *continuously releases cortisol, and chronic elevated levels can lead to serious issues.* Too much cortisol can suppress the immune system, increase blood pressure and sugar, decrease libido, produce acne, contribute to obesity and more.[16]

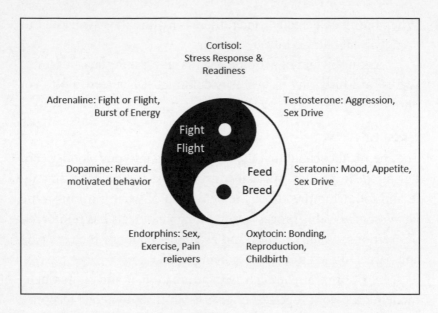

Is the stress reaction different for men and women? Well, sort of. Testosterone (the principal male sex hormone) and estrogen (the principal female sex hormone) play a role in how we react to stress, as do the neurotransmitters dopamine and serotonin.[17] For our ancient ancestors, dopamine may well have played an important role in reinforcing survival behaviors. And this mechanism may be part of our feeding (and, perhaps, over-feeding) behaviors today. This is explained well in a *Forbes* article:

> The smell of food stimulates the amygdala…a center
> of emotion, and causes further dopamine release…
> Sight, smell and taste of food stimulates release
> of endorphins (opioids) and dopamine…further
> stimulating the conscious part of the brain to eat...
> The human brain has a multitude of ways to stimu-
> late appetite and only a few to turn it off. That makes
> sense…because until recently mankind existed in a
> state of constant food scarcity. "Think about it: Your

brain is walking through these megastores and saying, 'Aren't I a great hunter? I can catch king salmon
or Kobe beef without any chance of being attacked by
a sabre-toothed tiger,'" says Mark Gold, distinguished
professor of neuroscience at the McKnight Brain
Institute at the University of Florida.[18]

Other scientists propose that dopamine is key to human
intelligence, including language and thought, memory and
reasoning.

But the big discovery in the research community lately is oxytocin.[19] Oxytocin is a hormone that decreases the stress responses,
reduces fearfulness, and enhances relaxation. It is present in both
males and females, though usually at larger concentrations in
females. It's an ancient and powerful hormone present in all mammals. It originates in the deepest part of our brain.[20]

Essentially, then, women's counterbalance to the fight-or-
flight response, to the stress of the hunt, is oxytocin…a feel-good
hormone that can produce euphoria. Men, on the other hand,
regulate through testosterone. Many people in the past have associated testosterone with aggressive behaviors. But newer research
indicates that testosterone is associated with status-enhancing
behavior, both aggressive and nonaggressive.[21] Testosterone is an
important factor in hunter-gatherer survival.

For both men and women, oxytocin is released under stress
and during sex. At the same time, while stress is relieved, the body
gets a shot of endorphins.[22] The main purpose of endorphins is
to reduce pain, but they may also produce euphoria. Many things
can cause their release, including eating chocolate, exercise, sex,
and *any situation that initiates the fight-or-flight response.*

> I think the core of my love of, my need for, hunting is found
> in the primitive, by which I mean ancestral, ecstasy felt at two
> moments…that moment of decision, that I am going to kill
> [at which time] I am as pure a creature as I'll ever be, involved
> in an act of monumental seriousness…And then I also hunt
> for the moment after the shot…a primitive sort of triumph
> in having killed…celebration and regret that leaves me awash
> with emotion, hyperaware of colors and scents and feeling
> physically lighter, as after extraordinary sex.
>
> —Bruce Woods, "The Hunting Problem" in *A Hunter's Heart*

Are You Feeling like Prey?

The term "fight-or-flight response" was first coined by Walter Cannon and popularized in his book, *The Wisdom of the Body*, published in 1932. Since then, stress researchers have added "freeze" to the possible responses. Some social scientists believe the "freeze" response is a stress reaction precipitated by a hopeless situation. Others believe that it is a moment of concealment (with no movement, it is hard for predators to see you) so you can consider your options and make a decision.

Dr. Peter Levine (medical biophysicist and psychologist) has a wonderful illustration of the freeze response and how very useful it can be. When Dr. Levine gives lectures on surviving trauma, he shows a video of a cheetah chasing a baby gazelle. He is trying to show the audience exactly how the freeze response works. He explains that the video is short because the average time for a cheetah attack from start to finish is about forty-five seconds. So, he makes this additional (and especially important) point about our stress response: it is designed to function for less than a minute. It's not designed to respond for weeks on end, as it is in chronically stressed humans who don't know how to turn it off.

In the video, the cheetah catches up to the gazelle. You watch as the cheetah ruthlessly sinks its teeth into the baby animal's neck. The cheetah throws it down on the ground several times, making sure it's completely immobilized. It's a game-over moment if there ever was one. Then something miraculous happens. When the cheetah gets distracted defending its "kill," the gazelle seemingly comes back to life. It's as if it is waking from a deep freeze. You see it shiver all over, and then it stands up and runs away, making a clean break.[23]

Chapter Two—Summary
Our ancient past and our modern life are inextricably intertwined. As the roots are to the tree (vast, interwoven, nourishing, uplifting, and usually unseen), so is hunting to humanity. We think of "going back to our roots" as seeking out our place of origin, revisiting the land where we grew up. But at a deeper, more profound, and more important level, it is a journey back to where we feel connected to the things that are important to life itself. It means returning to who we are. It is a pilgrimage to seek out our life source. And for humankind, that can only be found in hunting.

Chapter Three

"I'll Have What She's Having"

We tend to analyze our history from the perspective of present-day social conventions. In this chapter, we fight that trend. We will dig below the surface that our modern, urban way of thinking has constructed. We will look at why people hunt, even though most could easily feed themselves without hunting.

Many hunters have a difficult time expressing their motivation for hunting, so they use justification instead. When the media surveys hunters and asks, "Why do you hunt?" the hunter is likely to say, "Because it puts good, quality meat on my family's table." This is certainly a truly splendid outcome of the hunt, but is it really the reason we go out and endure the long hours, often in challenging conditions, in order to perhaps have a single opportunity to put meat on the table?

Is there a better answer? A more honest answer? A more complete answer? Yes. An anonymous user on an online forum could be the spokesperson for us all when he provides this honest and wide-ranging answer:

We hunt to be alone, to observe wildlife without being observed ourselves, to face one of the greatest challenges in this world: to take a wild animal on his own turf, using our brain and little else. We don't swagger into the woods and slay Bambi when he meekly peeks from behind a tree. We have to use every sense, every bit of experience we have, and when we accomplish our goal, it's a milestone. The adrenaline rush I get from it is like nothing else in this world. The fulfillment of long hard hours of hunting is definitely worth it![24]

It's in our blood, and it's in our bones. We hunt because that is what formed us.

The only adequate response to a being that lives obsessed with avoiding capture is to try to catch it…the sudden flurry of a partridge behind a thicket generates that strange contraction of our nervous and circulatory systems, whose symptoms seem extraordinarily like fright, although they represent the opposite of fear, since they end in an automatic movement of pursuit.

—José Ortega y Gasset, *Meditations on Hunting*

The "Predator Instinct"—Then and Now

Whether we are predator or prey, our bodies prepare us in much the same way. While researching the details of how our bodies respond in a predator-prey situation, the authors were struck by how similar the hunting brain-body reactions are to the sex brain-body reactions. As coauthor Dave said when we first discussed writing this book:

Man has three drives: to get food, to not be food, and to procreate. These drives are closely interwoven, and talking about things like this will help us to understand how much hunting is woven into our being.

The predator response derives from our ancient worldview. We categorize other living creatures as threats (predators), food (prey), or mates. If the brain shouts, "It's prey!" the cascade of chemicals invokes the appropriate actions to pursue, to capture, and to kill. And the consummation of this response is a strong euphoria and a healthy feeling of satisfaction. The hunter then brings meat back to the tribe, and his place in his social group is reinforced.

Dr. Paul Shepard, widely considered to be the father of modern deep ecology, advocated for a genuine understanding of man the hunter and a return to our hunter-gatherer roots. David Petersen supports Paul Shepard's point of view:

> Yet for an open-eyed segment of modern hunters,
> the ancient animistic sense of soulful—we might say
> "spiritual"—unity with wildness remains grounded as
> ever in awe, humility, and a strong sense of owed rec-
> iprocity. "Wildness is what I kill and eat," proclaimed
> the late Paul Shepard.[25]

Later in the same article, Petersen also says that Shepard is a "scientific philosopher" who says there's a gap between our original design and our current culture. And this gap results in the anguish and acting out we see in society today. He emphasizes that our culture is stressful instead of challenging. And he says we need the "magically *uncertain* adventure of the hunt."[26]

We, the authors of this book, believe there is a human hunting instinct. It's rooted in our basic survival structure that requires us to be a predator if we are to endure. It's exemplified in our brain-body response to stress. It's reflected in our brain's response to animals and their body language. It ties our hearts...no, our souls to the wildness that was our formative world.

The Third Driver

In the previous pages, we've discussed humans as predators (getting food) and prey (trying not to be food). Humans' next great drive is procreating.

You may wonder what this has to do with hunting. The most obvious connection is that if we are successful predators (and successful at avoiding being prey), we can stay alive, and we get to procreate.

But there's more! The brain chemicals that drive our ability to be predators and to avoid being prey are tightly tied to our sex drive. The same brain chemicals play a significant role in the act of procreation. In fact, from a biochemical perspective, the exhilaration of the kill is very similar to the euphoria of sex.

> We have gradually accepted human hands and legs as those of a hunter. Now we are ready to find in our heads the mind of the hunter: the development of human memory...[and the] moment of unity and freedom in ecstasy of intense release unknown to herbivores, based on the recognition of the perpetuation of life at the moment of the kill.
>
> —Paul Shepard, in the introduction to *Meditations on Hunting*

There's a famous line from a memorable scene in the 1989 movie *When Harry Met Sally*. In response to a bet, Sally, played by Meg Ryan, fakes a very strong and loud sexual experience while sitting at lunch in a deli. An older woman sitting nearby tells the waitress, "I'll have what she's having."

In fact, the oxytocin hormone we talked about earlier in the predator-prey response also plays a role in the sex response. While oxytocin is best known for the role it plays in childbirth and breastfeeding (it causes the appropriate muscles to contract), "animal studies have shown that it also has a role in pair bonding, mate guarding, and social memory."[27]

Since many of the same brain chemicals are in play during a predator-prey response as they are during a procreation situation, the body probably doesn't always distinguish the responses:

> Our control over sexual arousal is no better than our control over the dilation of our pupils or how much we sweat…It is a sign that our bodies react, just as they do with a rapid heartbeat or an adrenaline rush. We react.[28]

Hunter-Gatherer Sexuality

We have no way of knowing for sure how our ancient ancestors conducted themselves sexually, but we know that what they did lives on in us today. Some of the evidence is in our basic brain functions and the chemicals that our bodies produce. And we can make some guesses based on modern-day hunter-gatherer communities.

Modern-day hunter-gatherers (for example, the Sharanahua of South America) use meat to gain sexual access to wives and lovers. As Jared Diamond points out in his book *Why Is Sex Fun?*, there appears to be a status element here, with women generally preferring to mate with men who are good hunters.[29] And as described elsewhere, "across hunter-gatherers, the men most likely to have two wives tend to have higher status, [often] achieved through hunting success."[30]

The human female requires a certain amount of body fat in order to ovulate. This would make a good hunter an asset to her and to the community. And it may be that once the hunt is over and the community has enjoyed the feasting, both genders might be more relaxed…and ready to solicit and enjoy sex.

Some modern hunter-gatherer tribes apparently seek to enhance this impulse. They have a ritual of abstaining from sex prior to the hunt. Some tribes do this for a couple of days, and some do it for as long as a couple of weeks. Once the hunt is over and the hunters return to home base with plenty of meat, there is feasting and several days of unbridled sex.

Chapter Three—Summary

Why do we hunt? Most people could feed themselves without hunting, content to let others do their killing and harvesting. But hunting is an action that binds us to our most ancient past. It is the activity upon which the success of our species is founded.

Reproduction is another activity that is essential to our survival. Like most animals, humanity could perpetuate itself with annual breeding. But where's the fun in that? Something very similar could be said about how we chose to harvest our protein.

Fight or flight, feed and breed: these four functions are the definition of life. Fight or flight is for times of travail, a temporary state that cannot be sustained. But feed and breed are what we enjoy. They sustain and maintain us. They define our species.

In our past—and in our genes—sex and the successful hunt have always been intertwined. And the nexus of these two drives has formed our humanity. There is much to learn and much to be gained by understanding this essential interaction.

Chapter Four

What's That in Your Rearview Mirror?

What's the bond we have with our ancient ancestors? When we research and try to reconstruct their lives, we can feel the connection with them across thousands of years. They are our bloodline, and every day, in some way, we think, see, and behave just as they did. They did not know what kind of world we would be living in, but they prepared us as best they could.

The hunter instinct is an impulse from our ancient ancestors. It is not the only one. There are many things in our modern-day lives that resonate with their ancient lives in the predator-prey world. We'll discuss many of them in this chapter.

> The real hunter goes into that world of nature not as a casual onlooker, but as an active participant.
>
> —John Madson, "Why Men Hunt" in *A Hunter's Heart*

The Original Hunt Camp

So, at the end of the day, when everyone returns to the hunt camp, what happens? Of course, it depends on the camp, but most camps share some common features, which may be echoes of our distant past.

The evening is as noisy as the day was quiet. When the individual hunters end their day, they walk back to the hunt camp. Their muscles appreciate the stretching and the warmth of circulating blood. There's a quietness in the bush, and if the hunter is lucky, he'll walk home in moonlight. Perhaps he'll hear the call of a wolf…and the answer of another. An involuntary shiver…yes, the hunter is also the hunted.

At the hunt camp, the first one in stokes up the fire and lights the lamps. It's worth being late just to come in to this wonderful, glowing warmth. Someone starts the evening meal, and soon the smells of meat and onions fill the little cabin.

Once they have finished their chores, the hunters sit down to their supper. Sharing meat and telling tales. Completely ancestral. Oxytocin at its best.

> Although our attraction to wildlife predates the first time we ever held a gun…many of us have found hunting to be the most profound way for us to form a genuine relationship with wild animals and the nature they inhabit.
>
> —Thomas McIntyre, "What the Hunter Knows" in *A Hunter's Heart*

Shaping Anthropology…and More

Anthropologists have studied the current hunter-gatherer peoples of the world in an attempt to develop a better understanding of our roots. Of course, modern hunter-gatherer tribes are not the same as our ancestors. Indeed, they are not the same as their own ancestors. Time has passed, and needs have changed. We don't know what is traditional and what didn't survive "contact" with the outside world. But it is all that we have to study. We can only guess that there are similarities between ancient hunters and traditional hunters of the current era.

But there is one thing that binds the ancients to us, one thread that runs through thousands of years. It is the fear of predators. Fear of predators appears to be universal, and science is using this notion to inform research into many areas of study, including post-traumatic stress and childhood anxiety.[31]

So, while it may not be the whole story, the study of modern hunter-gatherers and the study of the primitive parts of our brains shed light on why modern people still have a hunting impulse and why it is critically important that we satisfy it.

> The cat hunts rats. The lion hunts antelopes. The sphex and other wasps hunt caterpillars and grubs. The spider hunts flies. The shark hunts smaller fish. The bird of prey hunts rabbits and doves. Thus, hunting goes on throughout almost all of the animal kingdom. There is hardly a class or phylum in which groups of hunting animals do not appear.
>
> —José Ortega y Gasset, *Meditations on Hunting*

Our Inheritance

There is a similarity among modern-day hunters that crosses cultures and creates bonds. It's easy for one hunter to join others no matter their country of origin, their color, or their faith. As hunters, we are united. We are one. We share the urge to hunt that comes from a deep center of our being. Our biology, our neurology, and our physiology are designed for hunting.

Carnivory

We are meat-eaters, and we got it from our ancestors. We are different from other high-functioning primates because of our protein-fueled brains. Our digestive system is different too; we don't have the gut capacity to process the high volume of vegetation that our bodies would require to extract enough nutrition.[32] Instead, our gut is built to process meat and especially cooked meat.

> Humans...consume a diet of much higher quality than is expected for our size and metabolic needs. This energy-rich diet appears to reflect an adaptation to the high metabolic cost of our large brain. Among primates, the relative proportion of resting metabolic energy used for brain metabolism is positively correlated with relative diet quality. Humans represent the positive extreme, having both a very high quality diet and a large brain that accounts for 20–25% of resting metabolism.[33]

> Others around me may opt to eat only plants, nuts, and fruits. Still others may employ faceless strangers to procure their meats, their leathers, their feathers, and all those niceties and necessities of life. Such is their right, of course, and I wish them well. All I ask in return is that no one begrudge me—and all of us who may answer the primordial stirrings within our hunter's souls—my right to do some of these things for myself.
>
> —M. R. James, "Dealing with Death" in *A Hunter's Heart*

Brains and Hormones

Hunting, and especially high-yield hunting, changed everything for our ancient ancestors—and consequently for us. That fabulous brain, with no equal in the rest of the animal kingdom, was made to hunt.

[Our ancestors] improved diet quality as indicated
by increased consumption of meat and bone marrow
and by tool-assisted food processing, at one point
including cooking.[34]

The brain functions that control our predator, prey, and procreation processes are linked. In fact, scientists have been studying
the brain's "reward circuit"…basically, the same brain areas and
brain chemicals that we've already discussed as the predator-prey
response.

And so, the body's design rewards us for every act of survival.

Is That My Reflection?

Vision

Color vision is an important part of the human experience.

As David Petersen asks in *Heartsblood*, did you ever wonder
why green is the universal color of restfulness and red universally
commands attention, excites, and agitates? Green is a color of
balance and harmony, of rest and restoration, of reassurance and
relaxation. It is used in art, homes, and institutions to provide a
gentle, calming environment.

Green strikes the eye in such a way as to require no
adjustment whatever and is, therefore, restful. Being
in the centre of the spectrum, it is the color of balance—a more important concept than many people
realise. When the world about us contains plenty of
green, this indicates the presence of water, and little
danger of famine, so we are reassured by green, on a
primitive level.[35]

Red is the color of urgency and importance, of courage and
strength, of fear and survival. We use it to make a specific item

stand out against a background, to emphasize its priority and ensure that it is noticed and dealt with.

> Being the longest wavelength, red is a powerful color. Although not technically the most visible, it has the property of appearing to be nearer than it is and therefore it grabs our attention first. Hence its effectiveness in traffic lights the world over. Its effect is physical; it stimulates us and raises the pulse rate, giving the impression that time is passing faster than it is.[36]

And red is the color of blood.

Processing Pictures

Human beings can process visuals very quickly. That was useful when our ancestors saw the tiger and got away from certain death. If they didn't see the tiger, they got to be dinner, and they didn't get to pass their genes along to us. The reason we process pictures quickly is because our brains process them simultaneously, whereas our brains process words sequentially…one by one.

That's why a picture is worth a thousand words. That is probably why cave drawings were an important part of our ancestors' lives. And our ancestors' ability to process visuals had a great deal to do with their survival…and our visual inheritance. (If our ancestors couldn't process images, and quickly, we probably wouldn't be here.)

> [W]e are genetically wired to respond differently to visuals than text. For example, humans have an innate fondness for images of wide, open landscapes, which evoke an instant sense of well-being and contentment. Psychologists hypothesize that this almost universal response stems from the years our ancestors spent on the savannas in Africa.[37]

Vision Skills

We mentioned earlier that humans have good vision, an effective mix of skills for color, near-sightedness, far-sightedness, peripheral, and binocular vision. We aren't particularly good at night vision compared to other predators. However, we are very fast at recognizing objects, and this may have saved us many times in the past. We are especially fast and accurate at recognizing animal images (compared to any other type of image and compared to the way other animals recognize them).

In a scientific assessment, researchers Holle Kirchner and Simon J. Thorpe were impressed by the "remarkable speed and efficiency with which humans process natural scenes." Using a test designed to have the participants move their eyes rapidly between points (saccade task), they showed that "when two scenes are simultaneously flashed in the left and right hemifields, human participants can reliably make saccades to the side containing an animal in as little as 120 ms [milliseconds]."[38] This rapid recognition was true for images of animals only and not true for urban scenes. The researchers estimated that performing the task in 120 milliseconds was probably as fast as the data could travel on the required neuron pathways. So, while they didn't think that our ability to perceive other animals quickly and accurately was wired in a special way, we are definitely faster at recognizing animals than any other object. Likely this ability helped us both as predators and when avoiding being another predator's lunch.

Context of Vision

Vision is affected by context. Hunters (and military snipers) learn to recognize the conditions that will compromise their ability to judge the size of and the distance to an object. For example, objects appear nearer than they are when they are clearly defined, in bright light, or viewed across low ground that's not all visible. And objects appear farther away when they are partly obscured (hidden behind light cover), in poor light (such as at dawn and dusk), or when the sun is in the observer's eyes. Ancient hunters most likely learned to accommodate these characteristics by hunting in daylight, by putting the sun behind them, and by choosing their ground for a clear view of the path to the prey.

Visual "distortions" also occur when we are stressed. One of the most common is tunnel vision, which occurs when someone is under extreme stress or threat. This would have riveted our ancestors' attention on, for example, a charging animal, which would have been extremely useful if he needed to time the use of a stone axe or a stabbing spear. In addition, many hunters report a slow-motion sensory distortion that may have helped the ancient hunter judge the distance and get the timing right. Another common distortion is an enhanced visual acuity of the surrounding area, something particularly useful when the hunter is dealing with a pack or herd whose individual members could charge or "buttonhook" the hunter. (Cape buffalo are renowned for looping around the hunter, or any predator, in order to attack from the flank or rear.)

> It is interesting to note that all of the sensory distortions…are extremely rare in normal life…*except* among hunters, where (for example) auditory exclusion is almost universal, and slow-motion time is very common. There may be something about the nature of hunting which taps into our ancient survival instincts. I believe that hunting is the only peacetime experience which will allow us to consistently tap into a "primal toolbox" of skills and experiences that is completely unknown to anyone else.[39]

Running

A great deal of study has gone into the way humans run. Physiologists have identified that the body structures that make running possible are different from those that make walking possible. This has led many people to speculate that our ancestors needed running skills to hunt (as well as, perhaps, to avoid being prey). This resulted in a theory that the hunting tactics of early man included "persistence hunting."

> Persistence hunting is a hunting technique in which hunters use a combination of running and tracking to pursue prey to the point of exhaustion. While humans can sweat to reduce body heat, their quadruped prey would need to slow from a gallop in order to pant. Today, [persistence hunting] is very rare and seen only in a few groups such as Kalahari Bushmen and the Tarahumara or Raramuri people of Northern Mexico. The technique requires endurance running—running many miles for extended periods of time. Among primates, endurance running is only seen in humans, and persistence hunting is thought to have been one of the earliest forms of human hunting.[40]

Anthropologist Daniel Lieberman and his colleagues noted that there appears to have been a long stretch of time when early humans had meat supplies without the use of arrows and spears.[41] Some anthropologists speculate that these early ancestors must have scavenged, while others speculate that they must have used other hunting techniques. Lieberman believes they used "endurance running," or "persistence hunting," as an important hunting tactic. What Lieberman adds to the discussion is a convincing analysis of what a man can and can't do, especially compared to his normal and usual prey:

> A deer and a decently fit man, Lieberman discovered, trot at almost an identical pace, but in order to accelerate, a deer goes anaerobic, while the man remains in an oxygenated jogging zone. The same is true for horses, antelopes, and a slew of other four-legged creatures. Since animals can run anaerobically only in short bursts before they must slow down to recover, a human in pursuit may have the final advantage. And because quadrupeds can't pant while they run, they also quickly overheat. To run down dinner, Lieberman realized, might simply have been a matter of spurring the poor beast into a sprint enough times to make it collapse from hyperthermia.
>
> "Running an animal to heatstroke is something that most humans can do, and that other animals can't," says Lieberman.[42]

Interestingly, the French word for "hunter" is *chasseur*, from the Old French, meaning "to chase."

Of course, there may well have been other hunting tactics that early hominids used. While scavenging and persistence hunting may have provided some delicious dinners, it's likely that they also used tactics such as ambush, stampeding, and driving over cliffs or into pits. In addition, they may have used tools that would leave no archeological evidence, such as sharpened sticks (plain or fire-hardened), snaring, lassoing, and netting.

We Are Hunters Still

We were all hunters then, and we are still, whether we hunt wild animals or pursue urban substitutes. You can see it in sport, in business, in sexual behavior, and in our everyday language.

In Sport

A significant number of sports teams are named for predators: eagles, raptors, hawks, grizzlies, lynx, panthers, sharks, rattlers, saber cats, lions, Bengal tigers, jaguars, bears, bruins, wolves, and coyotes, to name a few. And some just "cut to the chase" and call themselves "predators."

Most sports are based on various forms of hunting: throwing, running, intercepting, and so on. The most vividly hunting-related sports are things like the axe- and javelin-throws, archery, and their modern descendants riflery and shotgunning. But baseball's "hit-run-throw" skills are all hunting based, as are the many target sports like golf and curling. Even skiing, an early method of travel perhaps dating to 6000 BC, was originally developed to support hunters. Norse mythology describes the god Ullr and the goddess Skaði (god and goddess of the hunt) hunting on skis, and an early depiction on a runestone shows Ullr hunting with a bow on skis.

Modern nonadversarial team sports often use relays that reflect the techniques of persistence hunting, while adversarial team sports often demonstrate hunting techniques like baiting and ambushing. But most adversarial team sports involve the use of a notional prey animal (usually a ball, a puck, or a "birdie"), and the object is to get the prey into the net (or similar target area).

> The instinct of hunting and concealment is also ingrained in the human psyche, which explains why your children get such a thrill out of games of hide and seek.
>
> —Charles Stronge, *Kill Shot*

In Business and War

In cut-throat businesses, the predator-prey undercurrents are easy to see, whether it's external market behavior (where one company tries to dominate by "eating the competition alive") or internal to the structure of a business (where "only the strongest survive" and rise to the top levels of management).

The culture of modern business has been somewhat tempered by the entrance of women. In traditional societies, it is believed that women's skills (especially in social behaviors such as cooperation and reciprocity) shape the culture of the family or tribe. These traits have become more common and accepted in business as the number of women increases.

It is noteworthy, however, that arguably the strongest bonds and teamsmanship occur in military situations, which have, by far, been in groups of men (with apologies to the many fine female warriors of history who, while excellent, have been comparatively rare).

> Yet even among male groups, such as hunting parties and military units, the social emotions of bonding and reciprocity are extremely strong. In such cooperative forms of peer organization, the more "feminine" social emotions of trust, empathy, and reciprocity are required. Hence, the importance of oxytocin and serotonin rewards over adrenaline and testosterone... Even intense male activities such as combat and hunting are regulated by the social emotions of trust and empathy.[43]

The ability of humans to perform effectively as individuals and as members of the social group would have been critically important to early man. The individuals must each have had their own strengths in order to survive the harsh conditions, but they must certainly have worked as a strong team in order to hunt effectively and defend against predators. There is even evidence that older people were included at the family or tribal barbecue and were valued members of the social group, respected perhaps because of their knowledge and skills and their ability to pass these on to the younger, less experienced members of the tribe.

In Sexual Behavior

There's a large body of work and numerous competing theories as to the sexual behavior of our ancient ancestors. In general, the theories are largely speculative, and from one study to another, almost every type of mating system is proposed. As one team said in their study, the only thing that can be concluded by such studies is that we cannot come to any conclusion.[44]

We modern humans similarly cover most every mating system somewhere on this earth, though monogamy (or serial monogamy) and small harems are probably the most common. So, we can say that the roots of our mating systems are deep, and they often parallel those of the animals we hunted.

One of the charming characteristics that seems to be ancient and quite common among humans is the idea that the males should bring gifts to the females. Whether the gift is meat, sugar cane, or chocolates, in the offer of such items in the hope of receiving sexual favors, the motives and, frequently, the outcomes are indeed ancient.

Reflections in Language

We also hear the echo of our hunting roots in expressions found throughout modern language. Here are some examples:

- Lock horns – engage in conflict, from a fighting technique of horned and antlered species
- Rat race – competitive (but essentially pointless or fruitless) pursuit, from the prey behavior of rats in a captive state

- Corporate jungle – a place where there is an intense, ruthless competition for success, from a wildland where there is an intense struggle for survival
- Give it your best shot – making your best attempt, probably from the mid-1500 expression having "shot its spawn," a reference to fish fertilizing the females' eggs—recently appropriated by hunters using shotguns or rifles
- Job hunting – seeking new employment, from the stalking and capture of game
- Run with the hare and hunt with the hounds – to be a member of or to support two groups that are at odds with each other, from the predator and prey roles of hares and hounds
- That dog won't hunt – common Southern US expression meaning that what you say makes no sense, from the use of dogs to hunt raccoons or other wild animals
- Sniped, sniper – to attack from a place of hiding, from snipe (bird) hunting, which is normally done from behind cover

Anything, any excuse, to get out into the hills, away from the crowds, to live, if only for a few days, beyond the wall. That was the point of hunting.

—Edward Abbey, "Blood Sport" in *A Hunter's Heart*

Campfire Story: From the Cave
It was movement that caught his eye. The old man had said it would be. The antelope would soon be near, and he would have the chance to prove himself a hunter.

The boy held his spear in a throw position, waiting for the game to close within his range. He remained as still as the bush he was hiding behind. It was all happening just as the old man had said it would.

The old man had taught him how to throw, using a small shrub as a target. He started throwing at a range of only two steps. He threw only a few times, and his arm was sore. The old man showed the boy how and where the animals lived. He showed him how they would drink at the small ponds in one area and move to a wooded area to rest. After many years and many animals following this routine, the game trails were obvious, as were the favorite grassy areas where they would feed. The old man and the boy watched and noted the time of day when the game moved along these trails.

Then they would go back to throwing practice. The old man coached the boy's every throw, making sure each one was precise. Once the boy could hit the shrub with five consecutive throws, he would take one pace back and start again.

On another walk the old man took the boy to a cave near the village. Since the old man couldn't hunt anymore, he had more time to think about other things. He had wandered into the cave one day and realized he could make marks on the wall by scraping sharp stones over it. He thought of the time he spent hunting and tried to draw a picture to remind him of those glorious days. He drew many pictures showing the various ways he used to provide food for his family group. He discovered that certain minerals, when mixed just right with water, made various colors that he could use to enhance his drawing. He now brought the boy to see these drawings and explained the hunting techniques each one represented. Even a simple picture is worth a thousand words.

They had patterned the herd of antelope for many days and knew which of the trails they liked to follow and what time of day they would be there. They decided that early morning was best. The old man had placed the boy beside a camelthorn bush to provide concealment along the trail and give the boy room to throw his spear. When the antelope came along the trail, the throw would be about ten steps. The boy felt anxious; a lot was riding on the success of this throw.

An antelope moved along the trail as it had done most of its life. Suddenly it stopped and peered toward the camelthorn bush. The buck licked its nose, trying to improve its already superb sense of smell, but all it could smell was the rest of the herd following behind. The wind would not give away any secrets from behind the bush. The antelope's huge ears were straining for the slightest hint of noise that would instantly spark its flight reaction. It lowered and raised its head, trying to get a better view through the bush.

The boy watched through a small opening in the bush. The old man had told him how the prey animals had developed this sixth sense for danger and that the only way to overcome this was to avert his eyes. He was not to stare at the animal but watch it out of the corner of his eye or in short glances.

Finally, the buck switched its tail, as if to wave off the thought of danger, and turned its attention toward the lush grass waiting just a short distance beyond. As the buck began to move forward, the boy realized that this was it. He was going to get his chance to become a hunter. His heart raced, and he began to tremble. The old man had said this would happen, and that is why they had practiced on hare and guinea fowl to inoculate him to the excitement he would feel. He had learned to breathe slowly, deeply, and completely, making sure each breath filled his lungs and was fully expelled in turn. Soon he was calm and focused on what a good throw had to feel like. This was just another throw, just like the thousands of throws the old man had him do.

As the buck came even with the bush, the boy made his throw. It felt good, like so many other throws. He knew it would hit its mark even before it hit the buck. The buck had seen the movement and had tried to flee, but it was too late. The spear skimmed by a rib and sunk deeply into its lung, piercing it through and puncturing the second lung. As it ran off, the spear was dangling and flopping, creating more and more damage inside with each bound.

The boy looked back to where the old man had been hiding. He had left the boy on his own to kill his first antelope but had stayed near to help with the follow up. The boy now allowed himself to shake as he felt the glow of success. He wanted to immediately begin the follow up and claim his prize, but the old man had told him to wait until he was there to help. It took forever for the old man to get there although he was only a short distance away.

He calmed the boy and made him wait for a while—as long as it would take spit to dry in the sand—before they started to track the buck. They followed the tracks in the dirt a short distance until the blood trail started. The old man hung back and let the boy do the tracking. It was an easy trail to follow, and they soon came onto the downed buck. The buck was still alive and struggling to get up. The boy dashed forward and drove his stabbing spear completely through the chest of the animal. The buck quickly became still as if he were satisfied it was time to die. The boy again began to shake uncontrollably. He had accomplished his goal. He would now bring back food enough to feed his family group. Everyone would sing and dance with full stomachs. The hunters of the group would look on him with satisfaction and accept him as one of them. He was almost sad it was all over and that this fine buck had to die to give him all this.

The old man had told him about these feelings and nodded his approval. He reached down and wetted his fingers with the blood of the buck. He then carefully ran his bloody fingers down each side of the boy's face leaving behind red streaks of honor. The old man motioned toward the buck, and the boy knelt down and petted the antelope's head. "Thank you," he said. "Thank you for giving us your life. Thank you for giving us life...I will remember you for the rest of my life."

The old man nodded his approval and turned away so the boy wouldn't see the tears welling up in his eyes. "And thank you, boy, for letting me live my life again."

> The places we hunted are secret, known only to us and the deer. It would be heresy to share them with anyone. They are still there as I write this, in places so remote most would shake their heads, and in not too many years I will hike up there on bad knees and show them to my son. Those places— wild, beautiful, full of memories—are the greatest gift I could give him.
>
> —John A. Murray, "Climbing the Mountains after Deer" in *A Hunter's Heart*

"Why Do They Hate Us So Much?"
Why do some nonhunters feel such strong animosity toward hunters? Indeed, we have asked ourselves (as did David Petersen in *Heartsblood*), *Why does this incipient instinct to hunt flame so hot in some while seeming dead as cold ashes in most and even infuriating a few?*

For an abundance of reasons and circumstances, many people today are experiencing a complete disconnect from the past, from their own past. It's not that they don't know that modern man came from roots that run back in time through pre-industrialization, back to a predominantly agricultural time, back to Bronze Age and Stone Age, a time when we were all hunter-gatherers, and back beyond that. They know it intellectually, but they don't know it in their hearts and guts. They in no way identify with the ancient hunter-gatherer. They have severed themselves from that past. If closely questioned, they often believe the past is "in the past" and has no bearing on modern life. They believe the past might be interesting, perhaps, but not relevant. In the extreme version of this disconnect, some people disassociate themselves from the animal kingdom altogether. They believe that all predators should be fed manufactured food and should be prevented from hunting. And they think animal populations should be controlled by birth control rather than unregulated, "natural" birth-and-death cycles.

> José Ortega y Gasset asked, since humans' hunting is directly derived from our hominid ancestors' ways, at what time did humanity cross the line and step outside "nature?"
>
> —Chas S. Clifton, "The Hunter's Eucharist" in *A Hunter's Heart*

Most modern industrial people believe they are at the "top of the food chain." Of course, they know there are people in the world who are killed by animals—lions, tigers, elephants, buffalo, hippopotamuses, crocodiles, bears, and many more. But they prefer to think these killings are caused by circumstances, not by predators. These killings are exceptions to the rules. The animals' space has been invaded, the animal is "rogue," the animal has been baited, harassed, or otherwise annoyed by man. It's not that man is natural prey for these animals, they think; it's that man is in some way "bad" and deserves his comeuppance.

So, how did people come to think this way?

It's all about denial.

It's a denial that every one of us, in our current form, is an expression of the history of *Homo sapiens*.

It's a denial that all of us, no matter how educated and sophisticated, have the strengths (and weaknesses) of early man.

It's a denial that we are in the middle of the food chain, as we always were.

It's a denial that we are both predators and prey.

It's a denial of that dark space inside every one of us where we are killers and eaters of raw meat and that we have been fully formed by that drive and that ancient experience.

But maybe, more than anything, the main reason some antihunters are so sensitive is that they have capitulated as prey animals. Many members of our current society equate "prey" with "victim." And there are many examples of social and political groups who identify as victims.[45]

Similarly, and more dangerously, some antihunters equate "predator" with "oppressor." Hunters are predators. The prey will always have a deep, visceral fear of the predator. It may be the deepest, most primordial, and powerful of all fears. And fear is often the foundation of the most noxious and destructive hatred.

As we've said before, humans can be predator or prey, but it is no fun to be prey. We as a species are at our worst when we must run, hide, and live our lives in fear. Selfishness and lack of regard for others become the norm. The prey says, "I only need to outrun the slowest person in the tribe." The elderly and the weak are thrown to the wolves.

We can be at our best, noblest, and most generous when our tribe is filled with successful hunters. There are wonderful packets of condensed protein available for all! The poet, the thinker, the singer, the artist all thrive when the hunters are successful. There is room for the wisdom of the elders. They can be protected, and they can give to the youngers when the hunters are effective. The prey will leave the weak and old behind, but the successful predator will protect every member of the tribe.

As Kipling put it:

Now this is the Law of the Jungle—
as old and as true as the sky;
And the Wolf that shall keep it may prosper,
but the Wolf that shall break it must die.
As the creeper that girdles the tree-trunk,
the Law runneth forward and back—
For the strength of the Pack is the Wolf,
and the strength of the Wolf is the Pack.[46]

So, while some of the antihunters and most of the "neutral" parties understand intellectually that they are in the middle of the food chain, they don't dwell on it. They tend to think of the food chain as a pyramid, with "apex predators" holding all the power. They don't think of the food chain as being a sequence of predators and prey and that humans are prey when they look in one direction along the food chain and predators when they look the other way. As biologists note, just about every species, order, and class of animal includes both predators and prey. Mammals are no exception.

Most modern people, if they have ever felt like prey, have probably felt it from human predators. If they have ever felt heart-pounding fear, it most likely came from fear of a bully or a late-night stalker. Some people seek out a "recreational" activity where they challenge the environment in order to get in touch with that heart-pounding fear, like white-water rafting, sky-diving, or out-of-bounds skiing.

The hunter is grounded to the heartbeat of the earth by walking *as a participant* in the predator-prey world. When hunting deer, the hunter faces upwind and watches his six for a predator approaching from downwind. The hunter is intimately connected to life and death, and he likes it! The antihunter sees the life-and-death aspect of this and thinks it is all about killing, but it is much more than that. As one hunter put it, "It's the *intimacy* we love."

The antihunter is often filled with disgust, rage, and indignation…as if his own self were threatened. Psychologists say rage is sometimes triggered when we face in someone else what we most fear in ourselves. If that's right, the solutions for the rage of the antihunter are to be found within. His feelings of rage may be a pathology.

Every human fears death. But the hunter truly *knows* death and knows it as the antihunter does not. The hunter can find a degree of peace through his daily experiences with death. The prey will always fear death, and in fearing death, they may never truly live. The hunter knows death as an equal twin to life itself. Just as life and living define death, so do death and dying define life.

Hunters generally don't try to interfere with nonhunters: no protests, no bitter speeches, and no anti-vegetarian campaigns. Hunters go about their hunting ways and react only when they're poked. And that may be another reason antihunters can be so infuriated: hunters act like the lion who really doesn't care what its prey thinks.[47]

Chapter Four—Summary

The hunter instinct is a powerful impulse, passed on to us from our most ancient ancestors. The deep roots of our hunting past are not gone. The skills and physical traits needed to hunt are still visible in our bodies and their capabilities. The inherent predator-prey responses and instincts we developed in ancient times continue to provide valuable insights into human behavior in every aspect of modern life, from sports, business, and war to even our relationships and everyday language.

And we still have a hunting drive. The study of this hunter instinct sheds light on modern behavior. Some of us hunt our prey in the primordial forests, while others seek urban substitutes and modern surrogates like sports, business, and sexual behavior.

At the same time, some people in modern times manifest a strong animosity toward hunters. Could this antihunter behavior represent a dysfunctional disconnect from our roots? Like a bee that hates the hive, a beaver that is offended by the dam, or a wolf that rejects the pack, perhaps humans who hate the hunt are engaging in a pathological denial of their humanity.

There is room for both kinds in the modern world. The hunter does not attack those who choose not to hunt. It is reasonable for the hunter to ask for the same courtesy from those who do not respond to nature's wild call upon the human spirit.

In the next section, we'll look at the ethics of the hunter, then and now.

Section II

The Hunter's Oath

The hunter's ethics and ideals have a history, and their roots are tied to the practical and spiritual needs of man. Many of our modern hunting ethics started back in time when hunting was all about subsistence for every human on earth.

Our ancient ancestors were likely driven to improve their hunting abilities and tools in order to feed their families, whereas the modern hunter looks to improve his skills and technologies so he can increase the challenge of his hunt while still being sure of an ethical shot. Always, the hunter is trying to balance his skills with the abilities of the animal.

Chapter Five

Foundations

In this chapter, we look at the principles that underly hunting ethics. The first idea that we must explore is this: death is a part of life. We seek to understand death so that we can better understand life.

> The morality of the hunt is inseparable…
> from the spirituality of the hunt.
>
> —Theodore R. Vitali, "But They Can't Shoot Back: What Makes a Chase Fair" in *Hunting—Philosophy for Everyone*

Death Is a Part of Life

It is easy to romanticize life in the wilderness. We want to live in sync with the rhythms of the seasons. We want Mother Nature to nurture us. We see it as a joyous "off the grid" life. But it's harder to embrace the idea that nature's context is killing in order to live. Every organism exists in a cycle of life and death and life.

The "Food Network"

Man is not at the top of the food chain. In fact, the food chain is really a "food network." And man is somewhere in the middle of that. The largest predators as well as the smallest insects and bacteria all want to eat us and our by-products.

During our lives, if we live close to nature, we may have to defend our lives against bears, large cats, and other hungry beasts. The bush is also full of insects who take a little bit of us at a time. Even in our homes, there are dust mites who feast on our shed skin. And there are others who live right on us, cleaning up any nutrient they find.

And it's not just animals who are involved in this "food network." Plants also use the nutrients (the manure) we produce throughout our lives. And both plants and animals feast on our final nutrient gift in death. In fact, some naturalists and scientists refer to this cycle as converting plants to meat and meat to plants. So the meat of the white-tailed deer we hunt is a conversion of the plant life he eats. (White-tailed deer do eat more things than plants, but plants are the bulk of their diet.[48])

For many people, this "food network" is a repulsive thing. For them, the primary aim of interacting with nature is to see the birth-and-growth part of the "food network." They prefer to ignore or deny the need for death and nourishment. To naturalists and hunters, it is just a fact of life. For hunters, the primary aim of interacting with nature is to engage fully in the whole truth of life and death. The individual hunter's personal success is to get food and to avoid (or at least postpone) becoming food.

> Few wild animals die of old age. And their fate, their early death, is not cruel; it is simply the way of the wild. While some may call Mother Nature harsh, uncaring, savage, merciless, implacable, unrelenting—and worse—she defies all these humanistic accusations. Her world was around long before modern man, with his troubled conscience, stepped in.
>
> —M. R. James, "Dealing with Death" in *A Hunter's Heart*

Birth-Growth and Death-Nourishment

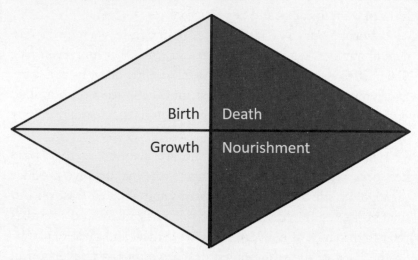

The struggle between these two views (birth-growth and death-nourishment) is relatively recent. It's likely that our early ancestors didn't struggle with these concepts. For early humans, these distinctions were irrelevant. The ancient human brain was designed specifically for getting food and for procreation. Those who were most successful at these tasks became our ancestors.

> Discomfiting as it may be to contemplate, we—as individuals and as a species—live because others die. This is a lesson that close, honest observation of nature teaches readily enough.
>
> —Mary Zeiss Stange, *Woman the Hunter*

The ancient human brain is responsible for essential functions. The ancient part controls breathing and blood flow, as well as vision, hearing, motor control, sleep cycles, and alertness. This part of the brain, along with hormones such as adrenaline (for energy bursts) and endorphins (for pain relief and pleasure), is essential for the predator to succeed in getting fed and procreating. Since the thoughts and behaviors of our early ancestors were governed by this ancient brain, hunting was likely natural and uncomplicated.

On Death

> Lions do not lie down with lambs in the wild,
> except when they are feeding off their carcasses.
>
> —Mary Zeiss Stange, *Woman the Hunter*

One of the key things we need to look at before we can start to understand the concept of the "food network" is the idea of death itself. It's hard to say when this concept formed in our ancestors' minds, but there are a couple of ways of looking at it. One is to look for archeological evidence of death ceremonies and rituals and then (without being too biased by our modern point of view) try to understand them from an ancient mind's perspective. For example, there are accounts of bones laid out in a path toward the cave where the bones of hundreds of animals lie. Were the bones in the cave and the bones marking the path intended as part of a ritual, or were the bones simply marking the path to what was used as a "skinning shed" or a kitchen? We cannot know, but our personal bias makes one seem more likely than the other.

Often we read accounts of ancient skulls and hominid bones being hacked or mutilated in some way. We can interpret this in many ways. If the mutilation occurred before the person died, it could have been done either as an attempt to provide life-saving surgery, as a punishment within a tribe, as an act of war, or as a part of cannibalism. (The authors vividly remember being told about the Australian aboriginal practice of breaking the legs of the human they were going to eat so that the meat would be kept fresh but the captive couldn't run away.[49]) And if the mutilation occurred after the person died, the reasons could be ritualistic in nature or attempts to make tools or crack bones to find any remaining marrow. It is difficult for modern minds to look at the evidence without a bias, and the bias we have causes us to think we're being logical when we're just being ourselves.

A Child's View of Death

What ancient man really thought about death is not known. It is likely that he accepted it as part of the way things were. He would have been exposed to death often—the death of animals in hunting and the death of people in his social group. When did hominids start to think about death as separate from life? A way to try to imagine what this process might have been like is to look at a modern child's gradual understanding of death:

- The first concept is that death is permanent. Toddlers don't understand that people who die aren't coming back. By the time children are about four, they start to realize that death is final.

- Later, a child starts to understand that dead people are no longer able to function as a live person. They don't feel, don't eat, don't dream. Children are usually between five and seven when they learn this.

- Finally, children grasp that death is universal. "Every living thing dies, every plant, every animal, every person. Each one of us will someday expire. Interestingly, before children learn this, many believe that there are certain groups of people who are protected from death, like teachers, parents, and themselves." Researchers say that children who understand that they will one day die believe it will be in some remote future.[50]

The most adamant of hunter-haters…are…moved, I think, by a fearful denial of mortality, especially their own…[They perceive] death as not merely unnecessary but criminally unjust.

—David Petersen, *Heartsblood*

Another View of Death and Killing

Another way to understand death—and therefore, killing—is to look at how modern hunter-gatherer tribes see it. Many modern tribes have legends and magic associated with understanding death. In many of these legends, death is seen only in terms of the "circle of life" (another name for the "food network"). The man becomes the animal he eats, and the animal becomes the man.

In other world views, and especially in mythology, there is a goddess who is responsible for birth and the hunt, so she represents the whole cycle of life from birth to death to rebirth. Many religions, both ancient and modern, include the idea of salvation and emphasize renewal, rebirth, or resurrection. And many people say that those who live well will die well and be rewarded with life everlasting in heaven.

In later chapters of this book, we'll look at killing and death in more detail. For now, we only want to say that how we view death may influence how we view the ethics of killing animals as a part of the hunt.

Life, Death, and the Hunt

Many critics of hunting base their objections on the belief that hunters are motivated by a twisted, single-minded pleasure derived from the act of killing an animal. No one familiar with hunting, and no credible sociological research, supports this notion.

—Richard K. Nelson, "Finding Common Ground"
in *A Hunter's Heart*

We would guess that our ancestors eventually had a philosophy of death. Once we are capable of understanding death, life changes. In the history of humanity, when did this occur? Tool development may have driven all other forms of development. Man's use of tools (for hunting and probably for personal defense) goes back to our earliest beginnings. From one powerful perspective, tools define who we are. While it may be hard for us to picture from our modern perspective, it is possible that ethics originated early on in our ancestors' development. And the ethics might have developed out of the following basic ideas.

Principle One: Conservation of Energy

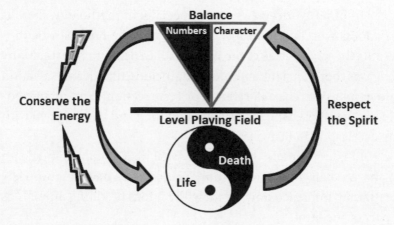

Protein ≥ Effort + Risk

The protein you get must be equal to or greater than the price you paid. The price includes both the effort and the risk. If there is risk, the protein must be worth it. This is the foundation of predator-prey relationships. The effort expended (and the risk taken) by the predator must be proportionate to the food value of the prey. It's easy to see this principle at work in modern hunter-gatherer tribes, where a large portion of protein is acquired through insects, grubs, and small animals that are available daily with little or no risk.

Other carnivores and omnivores see it the same way. Grizzly bears eat the highest-calorie part of the salmon (skin, brains, and eggs) and often toss those beautiful pink salmon steaks back into the water. While a wolf pack will seek out the newborn deer and moose in the spring, they'll also eat mice, voles, and other small game throughout the year. Chimpanzees are omnivores and will eat not only leaves and fruit but also termites and animals such as bush pigs and antelope. They hunt alone and in large parties, where they effectively drive the game (often, colobus monkeys) into an ambush of waiting chimps. They also practice cannibalism and have been known to steal and eat human babies. We find this particularly interesting because chimpanzees are our nearest relatives, and yet humans are "fair game."[51]

Our modern human sensibilities may find these chimpanzee behaviors revolting. But it's completely acceptable from the point of view of predator-prey energy dynamics. If the prey is available and takes only a relatively small amount of energy to acquire, it is "good" for the predator to take the prey.

Modern Hunting Efficiency
When we talk about modern hunting ethics, the idea of the conservation of energy is applied a little differently. Today, few of us are subsistence hunters, and our lives are not dependent on the success of the hunt. So hunting efficiency has been superseded by other hunting ethics. Modern hunting has evolved to be far more ceremonial than "fill the freezer" requires.

In fact, hunting is not the most efficient means of putting meat on the table. Agriculture is. The rapid expansion of the human population over the past ten thousand years has many contributing factors, and one of the key factors is the efficient production of large quantities of food through the domestication of animals and plants. If the modern hunter were only interested in the most efficient way to put protein in his stomach, he would choose farming…or the supermarket.

Nothing Wasted
However, this principle of energy efficiency is probably the basis of one of our most cherished hunting ethics: use every part of the

animal. Nothing wasted; all parts treasured. We make sure that everything edible gets to the table (if not our own table, then our dogs' bowls) and that the animal trophy (inedible for us) is brought in from the field and given a place of honor in our home. This is an echo from our past, where the pelt would have been used for clothing and the bones would have been used for tools. More recent ancestors would have given a small token of the animal back to the land or the gods as thanksgiving, a tradition still followed by many modern hunters.

Sometimes modern hunters feel it's a little too easy to fill the freezer, so they structure their equipment or their hunting techniques to *decrease* the efficiency of their hunt. Why would they do this? One reason is that most hunters enjoy the process of hunting, and making the hunt less efficient (and therefore more challenging) prolongs the time they spend in the wilderness. Another reason is that they believe that by ensuring the hunt is on a level playing field, it is ethical.

Principle Two: A Level Playing Field

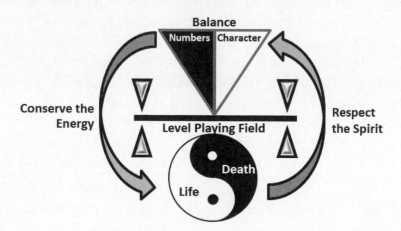

Challenge

Modern technology can improve our ability to take game more easily. Certainly, the emergence of tools helped move humans up the food chain. Tools helped our ancestors be successful. They went home empty-handed less often. They more often got protein proportionate to the energy expended.

There are modern hunters who use spears and traditional longbows that mimic the hunts of our ancestors. However, most of the hunting arms currently in use (compound bows, crossbows, handguns, black powder rifles, and modern rifles) give the human hunter an advantage over the ancient arms. The advantage is not so great as nonhunters imagine. It is mostly an advantage of distance. We can more likely successfully "sneak up" on a deer if we only need to be within one hundred yards to shoot it with a rifle, whereas with a bow range of about thirty-five yards, we are more likely to have to lure or ambush it.

To level the playing field while using modern hunting arms, hunters use other restrictions to make the hunt more difficult. We follow a game-management program that allows us only to take a specific gender and/or age of animal and only within a limited time period (from a few days to a few weeks each year and usually only during daylight hours). There are many laws about hunting, each tuned to the requirements of their jurisdiction, and most of them have to do with using hunters to manage game populations.

In addition, many hunters add to the challenge by making their prey harder to find. Some hunters will purposely hunt in difficult terrain, or they choose the late season when it's the coldest, or they only hunt by stalking. Other hunters will take only the oldest animals or the animals with the largest or most unusual antlers. A few hunters will hunt only predators (or dangerous game), and this adds to the hunter's personal risk. These things all increase the challenge for the hunter and level the playing field for the game.

> Some of us, upon achieving greater skills at taking our game…begin to handicap ourselves at some point in order to foster greater intensity of experience, and thus greater intimacy between ourselves and the animal hunted.
>
> —Theodore R. Vitali, "But They Can't Shoot Back: What Makes a Chase Fair" in *Hunting—Philosophy for Everyone*

Risk

Every hunt includes some risk to the hunter, some more than others. Because humans are somewhere in the middle of the food network, there are predators who would like to kill us. Modern city dwellers don't encounter these creatures except in zoos and in the media, but in the wilderness and the jungles, they are real, and they aren't anything like the Disney cartoons. At the top of the list of animals that like to eat us "soft gooey ones" are lions, tigers, brown bears (including grizzlies), wolves, sloth bears, hyenas, piranha, sharks, crocodiles, leopards, cougars, Komodo dragons, and alligators.

From a risk perspective, there are also animals that will attack if they feel threatened (or grumpy) even though they don't intend to eat us. Among these are "dangerous game" animals such as Cape buffalo, hippopotamus, elephant, and rhinoceros. It is very expensive to hunt these animals (when it is allowed at all), and there is no doubt that the risk is high.[52] And even animals that aren't normally considered dangerous game can add to the hunter's risk because they can be very dangerous during their mating and birthing seasons. North American hunters ensure they don't get between a mother and babies of any species, and they are always cautious about getting too close to horns and hooves during the rut. Wilderness hunting brings its own risks of injury caused by terrain and weather. Furthermore, the distance to medical help may turn a small problem into a dangerous situation.

> Further, the nature hunter needs a credible challenge: to work hard, to suffer and sacrifice at least a little, and to fail more often than succeed. By approaching hunting in this adventurous spirit, when I do bring home some meat I know I've earned it, both morally and physically.
>
> —David Petersen, *Heartsblood*

Principle Three: Balancing the Numbers

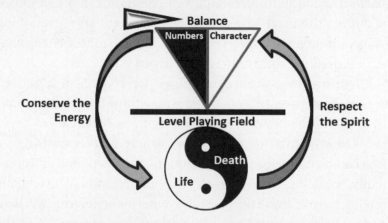

Another aspect of the hunt is that we must balance the numbers of predators and prey in each area. This is not a modern idea. Our ancestors would move around to new hunting grounds. This would increase the number of prey available for their hunts. As the prey got thinned out, hunting became more difficult and time-consuming. That would be a good time for the ancient hunter to move on to another area. For the modern hunter, there are many rules concerning what we can hunt and kill, with the oversight of governments to develop and manage the game plan to balance the numbers.

Genetic Empire-Building
The whole point of life is to further our genetics. Although modern people often think there are many other points to life, the fact is that if our ancestors hadn't furthered their genetics, we wouldn't be here to argue the point.

All animals perpetuate their species by procreating. To procreate in large numbers, animals need habitat that includes good sources of food. This simple fact drives humans not only to eat and multiply but (since we are *sapiens*, or "wise") also to plan where our next meal is coming from.

The foragers of ancient times hunted in a sparsely populated world and moved around from one hunt camp to another, not putting too much pressure on any one area. Then everything changed. It took a while to get started, but the rate of change has been accelerating since humans first appeared.

During the so-called Cognitive Revolution, our earliest ancestors may have developed new ways of thinking and communicating, and they started to develop cultures that enabled larger groups to work together for hunting, for war, and for trade.[53]

Later, an "Agricultural Revolution" provided food as a commodity, enabling rapid growth in populations and eventually enabling most of those people to live in permanent farms, villages, and cities. This resulted in distinguishing "wilderness" from the rest of the world.

The Scientific Revolution (usually dated from the time of Copernicus) started well less than one thousand years ago. It was followed closely on by the Industrial Revolution, which not only changed the ways of industry but also revolutionized agriculture. This changed life as we know it.

It's now a vastly different world. The human species, over seven billion and climbing rapidly, is arguably "out of balance" by the numbers. We are now the dominant species. We're not dominant in terms of total numbers of animals or biomass or territory. We are only 7.7 billion people (as of this writing) compared to a total estimated world population of twenty quintillion animals,[54] most of which are insects, birds, and fish. For each *one* of us, there are about three billion other animals sharing this earth. We're not the most pervasive. Current estimates are that we humans inhabit about 1 to 3 percent of our planet. But we have a huge influence on what happens in this world.[55]

Balancing the Impact of Hunters

The good news is that most of the people in the world do not hunt and don't need to rely on hunting to feed themselves. Indeed, the world population in 5000 BC was likely about five million. The majority were still hunter-gatherers, and there were lots of animals to go around. But the agricultural and industrial revolutions enabled huge population growth, and we can't now go back. If we all became hunter-gatherers overnight, we'd soon consume every animal on the planet.

So, it's about balance.

All hunters want to hunt, and they want to do it at a level that sustains the population of game animals. As predators with a conscience, we regulate (and pay for) the hunting process to ensure that happens.

Within the bounds of available habitat and nourishment, it may be that prey have many babies in order to provide enough food for predators. In nature, there does seem to be a balance of predators to prey to habitat. Most hunters are very aware of this. There are years when partridges and rabbits increase in number, followed by years when coyotes, wolves, and foxes increase in number. This repeats at intervals over time.

Ensuring the Survival of the Wild Animals

Hunters respect the need to guarantee the survival of game animals in enough numbers to safeguard the survival of the species and the local populations.

At the time of European contact in North America, there was little regard for conserving wildlife or habitat. The natural resources (primarily trees and fur-bearing animals) were harvested and shipped overseas with no thought for making it a sustainable economic activity. Commercial meat hunting and subsistence hunting were not regulated, and many animals were indiscriminately killed.

A group of men in 1887 formed the Boone and Crockett Club, which began keeping records of big game kills, particularly those animals the men worried would become extinct. According to an interview with Jack Reneau, former director of Boone and Crockett Big Game Records, "A scoring system wasn't to give sportsmen…bragging rights." The original purpose of the records was to document the animals living in North America and to save that information for future generations.[56]

The outstanding success of the deer management programs of the past one hundred years has turned game management upside down. The original purpose was to restore the deer populations. Now, there are lots of deer, which creates conflicts with hunters, farmers, property owners, vehicle drivers, and city-folks. The wildlife manager needs to work with these groups to devise effective deer management strategies.

Hunters are very conscientious about making certain their hunt fully supports game management. The government designs the licensing and tagging systems so that only suitable proportions of genders and ages of game are taken. But more than that, there is a broadly subscribed hunting ethic to take only the oldest members of the pack or the herd, especially those past their prime for reproduction and likely heading toward their last winter or their last dry season.

There's an excellent example of this in a rhino hunt in Namibia. Hunters participated in an auction to obtain a permit to hunt black rhino bulls that scientists had determined to be hazards to the survival of their own species. As the bulls get old and are no longer able to breed, they become aggressive and may even charge and kill younger bulls, cows, and even calves. When an old bull is among a herd, it increases the herd's mortality and reduces its productivity. The goal of removing the bull benefits not only the herd but also the Namibian conservation efforts. The winning bid of $350,000 one hunter paid for the permit went to the Namibia government, which planned to use it to help fund anti-poaching efforts.[57]

Principle Four: Balancing the Character

> The caribou feeds the wolf,
> but it is the wolf who keeps the caribou strong.
>
> —Inuit Proverb

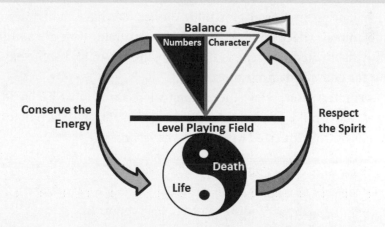

When it comes to predators and prey, one defines the other. If the antelope were slow, the cheetah would not need to be fast. If the wolf were not stealthy, the deer would not need to have superb hearing and sense of smell. (And of course, the converse is true: if the deer were less attentive, the wolf would not need to be so stealthy.)

There is a very tight relationship between predators and prey, not just in the sense that one is committed to feeding the other. Everything that animals are has been designed by the predators they face and the food they consume. Even humans were formed by the predators they faced and the food they hunted or gathered.

In our very recent history, humankind hasn't had a direct effect on the genetic design of wild animals although we certainly have influenced their habitat (and that could eventually affect them genetically). But we've had a huge direct effect on domestic animals. In modern times, through selective breeding, we have "designed" dogs, cats, horses, cattle, pigs, sheep, goats, rabbits, and all sorts of poultry. Humans, having domesticated the animals they like to have the most of (pets and food animals), no longer have very much direct interaction with wildlife.

However, we've affected the *behavior* of many animals. We now have raccoons who are completely habituated to urban life and reliant on dumpsters for their dinner. We have many large game animals (deer, moose, and elk, for example) who are sufficiently habituated to humans that they seek urban and suburban areas for both food and protection.

Hunters don't view these habituated animals to be "wild game" and would not consider hunting them except as part of animal control requested by the local government.

The pleasure of the sportsman in the chase is measured by the intelligence of the game and its capacity to elude pursuit and in the labor involved in the capture.

—John Dean Caton (Chief Justice of the Illinois Supreme Court), *The Antelope and Deer of America*

Corollary to Principle Four: Animals Are Not Subhumans

Many modern people like to think of animals as possessing human qualities. In fact, they often regard them as underdeveloped humans. From one perspective, this is human arrogance at its height. The hunter would say that, in order to deeply appreciate them, we must view each animal as a fully formed creature, not some sort of mini-human or subhuman.

Predator animals aren't so cute, but people like them when they're still babies. Babies don't kill; they nurse or eat what the adults bring. Baby predators are adorable…lion cubs, wolf pups, bear cubs…all adorable. We even put toy versions in our own babies' playpens.

Since we are human-centric, we tend to humanize animals. We might learn something if we tried being animal-centric for a while. What if we looked at ourselves as animals? Imagine for a moment that the human animal is driven first by the need for individual survival. Our priorities are sustenance, shelter, and security. Then there's the need for our genetic survival: sex, birthing, and the care of young. Any of the components of individual or genetic survival may be better assured by some form of social group: a family, a clan, a tribe.

> It is both erroneous and wrong philosophically, I think, to assume that all creatures want to be men or that human consciousness is the cosmic pass toward and through which all life must make its way.
>
> —Paul Shepard, *The Tender Carnivore and the Sacred Game*

So, then babies are "good" because they help our genetic survival and "bad" because they threaten our individual survival by being needy and noisy enough to attract a predator. But genetic survival is the only way we'd be here to have this conversation, so genetic survival is more important to nature than is individual survival. This last perspective—that genetic survival is more important than individual survival—is fundamental to game management.

Animals and Environmental Change

In terms of ecological and animal impact, modern hunting is a lightweight. Habitat destruction and environmental changes resulting from population expansion have, by far, a greater impact. Even the exponential rise in our standard of living, our own personal environmental footprint, would not be significant were it not for the world population growth. The impact of habitat loss is at least an order of magnitude greater than climate change, as the climate has always been changing. The Ice Age has come and gone, but within the last two hundred years, there has been a phenomenal loss of the truly wild spaces.

If the environment changes, the animals who can adapt and capitalize will survive and thrive. We lost the dinosaurs about sixty-six million years ago because the environment changed and they could not adapt. The most adaptable creatures today are probably cockroaches, who can thrive almost anywhere and have been on this planet for at least 320 million years.

Principle Five: Respecting the Spirit

Indigenous people often best expressed respect for the spirit of the animal. As Jacob Wawatie and Stephanie Pyne wrote, they "understood themselves as predators," profoundly connected to the prey. They understood that their lives "depended on taking a life." And they were grateful to the prey who sacrificed themselves. Therefore, they say, "the traditional way was to hold feast ceremonies in honor of the prey, and to practice an ethic of not wasting anything from it."[58]

Among traditional peoples, animism (the belief that everything in the universe has a spirit) is common. They believe the

Great Spirit is the supreme being. The Shaman mediates between the earthly and the spirit worlds. The Animal Spirit, or spirit guide, walks through life with a person, teaching and guiding them and, in some instances, protecting them.

One of the common themes in Indigenous North American stories is the idea that an animal specifically gives permission to have the members of his herd hunted. The spirit may send the animals down the mountain or the fish up the stream. Or the animal, by appearing before a hunter, gives permission to be killed. The spirit of the animal may pass to the hunter: "You are what you eat." Or the spirit might be reborn in another animal.[59]

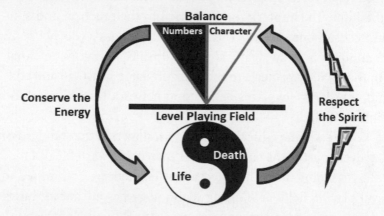

The hunter who spends long, lonely, and lovely hours in the wild can easily understand the profound spiritual aspects of hunting. When you sit quietly and hear the whisper of the wind, you know there is life in the trees. You watch your hunting area, and suddenly a deer appears, not walking into your view, but already standing there when you first see it...and it's easy to believe the animal came to you knowing you were there to hunt it. When you're walking in the wild and you come across the rich smell of black loam, the acrid odor of wet autumn leaves, or the overwhelming sour smell of a black bear, you know you are touching the spirit that energizes all of life. When you hunt and your deer is down, you watch the life drain from its eyes. The mystery of life and death, the greatest of *all* mysteries, flickers before your eyes. And when we touch its still-warm body, there is something in most of us that knows a great spirit has given itself to us.

This is a consistent theme in hunting, across the ages and across the globe: hunters honor the animals. The hunter understands that, from one critical, essential perspective, we are all spirits under the same skies.

Chapter Five—Summary

There is a powerful historical foundation for the ethics and ideals of the hunt. The hunter's roots are tied to the practical and spiritual needs of our species. The hunter understands (deeply, fully, and completely) that death is a part of life. Every living thing, from the smallest bacteria to the apex predator, must kill and consume in order to live. This is the food network that sustains and maintains all life.

Hunters value ethical efficiency, as shown in the standard of using every part of the animal rather than being wasteful and thus leading to killing more than necessary. Hunters are committed to a level playing field: as hunting technology has advanced, hunters apply greater regulations to increase the difficulty of the hunt. Some hunters even go beyond the legislation and create further personal restrictions to increase challenge or risk. They responsibly contribute to game management and show a deep respect for every animal.

These hunting ethics developed from practical roots, which in turn better equipped hunters to feed themselves, their families, and their tribe. The principles and ethics of hunting sprang forth from motives of balance and respect. They are the end product of a lifestyle suited to sustain the food network…the cycle of life.

Chapter Six

Killing and Connecting

In the beginning, the hunt was literally a way of living. It was a way to find food. Most animals don't distinguish between hunting and killing. Animals hunt, catch, and eat. If need be, they immobilize while catching or while eating.

After a vast amount of time living by hunting and gathering, when humans became farmers, wild animals became "pests." And at least to some degree, hunters became pests too. Gradually the farmers started to control their land and its boundaries. Settlements grew. Royalty or civilian authorities controlled the land and its inhabitants, including game animals. Unwanted hunters came to be called "poachers."

In this chapter, we try to define what hunting is now that its context is civilization and modernity.

That Defining Moment

For many hunters, the hunt is a vital, visceral activity that throws one both physically and mentally into the heart of the natural world and confronts one with the fundamental issues of existence. The hunt, for them, is no simple recreational activity but a fundamental celebration of the natural world and life itself.

—T. R. Kover, "Flesh, Death, and Tofu: Hunters, Vegetarians, and Carnal Knowledge" in *Hunting—Philosophy for Everyone*

Hunting for Beginnings

Our ancestors fed themselves by eating what was at hand and moving to where there was more of it. If there were too many creatures trying to eat them, they would move on to a place where the balance was more in their favor. They persisted in this way for thousands of years. During the Stone Age, they probably started to think in terms of foraging to obtain food: scavenging, hunting, and gathering. Each of these would require its own tools. With the advent of cooking, scavenged food became healthier to eat. As well, cooked meat and plants were easier to digest.

At this time, *Homo sapiens* began to thrive. Having spent thousands of years developing our cultures and our skills, we started to make leaps in our development in much shorter time periods. From our early days, we can find evidence of symbolic thought, language, finely made tools, games, music, and burial.

In the time of early literacy, perhaps as late as the beginning of Genesis (estimated to be about 4400 BC), man started to move in large numbers from hunting to farming. Certainly, by that time, man would have defined "hunting" as separate from farming. In fact, there are scholars who believe the Adam and Eve story in the Bible captures that moment—that the expulsion from the garden of Eden represents the passage of *Homo sapiens* from the hunter-gatherer to the farmer. This passage is sometimes called "the other side of Eden."[60]

From a hunting perspective, this passage from Eden is the defining moment for hunters. Until that time, there was no other world but the wild one and no other means of finding food than for-aging and hunting. In his essay, "The Sacred Pursuit," philosopher Roger Scruton describes how some writers portray our passage from Eden, at which time we domesticated animals and we also became a domesticated animal. He believes that our sense of Eden is in the "deeply buried layers of the human psyche." Scruton writes:

> The native Americans described by [James Fenimore] Cooper [in *The Last of the Mohicans*] are no longer hunter-gatherers. But the book contains incompa-rable descriptions of the other side of Eden—how it looked, sounded, and smelled, before man the farmer exterminated his hunting brothers.[61]

Whether it was a physical or a cultural extermination, hunt-ing became a nonessential activity. Agriculture was intended to feed everyone. Hunting became a social activity, one usually allowed only to the elite class. Game parks (reserves or preserves) provided game-rich areas owned by and hunted by the upper class and royalty.

But hunting was not a royal sport in the United States. When Europeans made contact, the Indigenous peoples of North America were mostly hunter-gatherers. Formed in revolution, the United States was proud of not being ruled by nobility and aristocracy. As James Carmine writes in *Hunting—Philosophy for Everyone*:

> American game laws grew in strident opposition to the tragic consequences of the enclosure laws of sev-enteenth- and eighteenth-century England. These en-

closure laws eliminated access by the common people
to common land and thereby forced them off of the
land upon which their lives depended...All men are
created equal, and [in the United States] the common-
er is as free to hunt as the aristocrat or the gentry.[62]

While only a small percentage of American people hunt
(usually estimated at about 5 or 6 percent), many Americans
support hunting. Survey results reported by the *Pittsburgh Post-
Gazette* in 2007 found that 80 percent of respondents indicated
that hunting has a legitimate place in modern society.[63] A more
recent survey (2019) by Responsive Management and the National
Shooting Sports Foundation found similar results.[64] One of the
effects of the COVID-19 pandemic appears to have been a significant increase in those numbers.[65]

Many people in the United States no longer know the history of European aristocracy and the importance of American
common lands in this history. However, the majority appear to
understand that it has something to do with freedom, and that's
why it's so defiantly cherished.

It's a symbol of the common folks having access to game and
game lands. And here we have it: the hunt is now, more than ever,
symbolic. Thus, a true understanding of hunting needs to address
all these elements: physical nourishment, psychic nourishment,
social nourishment, and symbolic nourishment. To paraphrase
George Orwell, that trophy on the wall of the common man's home
is a symbol of democracy. It's our job to see that it stays there.

What's Hunting?
Hunting is searching for game or other wild animals for the purpose of killing.

The underlying truth of the hunt is that death is a part of
life. Humans' original state is to include meat in our diets, and the

original, most natural way to do that is to hunt. Humans are predators, and predators connect with the natural world in the kill.

> How can anyone who hasn't seen and touched death know or understand?…I know, in fact, that they cannot…Each successive sheltered generation in turn widens the growing chasm between man and the land.
>
> —M. R. James, "Dealing with Death" in *A Hunter's Heart*

Just as some predators are incited by the prey running away and instinctively give chase, so humans instinctively hunt.[66]

During the hunt, the hunter may start to feel he is getting truly in touch with his mid-brain, the part of his brain that speaks to vision and perception, fight and flight, feeding and breeding—that's his "hunting brain." At the moment of the kill, the body receives a shock of hormones. This can cause uncontrollable shaking and an urge to do something physical. In that instant, the hunter may feel a swell of sadness at the moment of the kill. The hunter connects deeply with his Paleolithic ancestors in a shared act of continued survival.

> In order for there to be a hunt, killing must be the goal: it's the instinctive spark that ignites the ancient predatory fire within all true hunters.
>
> —David Petersen, *Heartsblood*

Death and Denial

The hunter, by his very acts, cannot deny mortality…not the game's and not his own.

New hunters often wonder how they will feel when faced with the death of their animal. We tell them there will be a moment of connectedness with the animal where they will feel a pang of sorrow. The animal has given its life for this moment. But as the life drains from its eyes, its heart stops, and it breathes no more. It quickly passes from being an animal to being meat.

It is at that moment everything about life suddenly comes into focus. We are here because our ancestors long ago did the same thing we just did. Successful predators, we are united across untold eons.

For the true hunter, there is death, but there is no denial.

> My reindeer lay at my feet with its head on the ground, mouth open, and a pool of blood forming. I could smell it. He lived a little longer, watching me, knowing at last who was the stronger of us.
>
> Ah well. Our eyes met, and I held his gaze until his pupils widened in an empty stare. Then I was alone, with meat.
>
> —Mary Zeiss Stange (quoting Thomas's *The Animal Wife*), *Woman the Hunter*

From Subsistence to Slaughter

Hunting by Any Other Name

A lot of terms are used for hunting. You've probably heard many of them like "subsistence hunting," "sport hunting," "ranch hunting," "trophy hunting," and more. These terms are often used divisively. We will use them only where the descriptor adds clarity. As much as possible, we'll call hunting "hunting" and leave it at that.

There are also a lot of definitions of hunting. Many of them attempt to exclude specific types of hunting. Again, we're not going to go there. Our goal should be the broadest possible definition of hunting that captures the spirit of it, regardless of who is doing it or how they're doing it.

So, here it is: hunting is searching for game or other wild animals for the purpose of killing.

Now, here's what it doesn't include:

- Killing domesticated or tame animals
- Killing any animals that are tethered or otherwise disabled from escaping
- Scavenging (finding game already killed)
- Searching for any game or wild animal without intending to kill it

So, slaughterhouse killing is not hunting, and photo safaris are not hunting. Broadly speaking, trapping and fishing are hunting. Poaching is hunting, but poaching is illegal.

Agriculture Made Us Wild

The advent of agriculture defined "hunting." Before agriculture, there was simply feeding ourselves from the wild. Agriculture changed the game in many ways. Land became property. Property became "owned." Hunting, for the most part, was no longer the primary means of subsistence. Farming took over the job of feeding people, and the hunter-gatherer was marginalized. Nowadays, most of the few remaining hunter-gatherers live in areas where farming is not practical, such as the arctic and the desert.

As agriculture gained strength, domestic animals started to feed everyone. And farmers killed off wild animals. They killed any animal that put their herds and flocks in danger, and they tried to exterminate those that ate their crops.

The idea of "varmints" and varmint hunting arose. Varmints were mostly the smaller wild animals that may have provided a mainstay in the diets of hunter-gatherers. In some cases, the new ecology of agriculture inadvertently and unintentionally favored a particular animal. That animal flourished and became a "pest." The rabbits in the vegetable garden, the rats in the granary, the fox in the henhouse, the birds in the orchards…all became pests. The most recent animal to be classified as a pest in North America is the European wild boar (and the feral pig and hybrids). Capable of ruining acres of crops overnight, wild boars also dig up the ground, endangering cattle and horses. (The rutted ground can easily break their legs, resulting in them needing to be euthanized.)

Farmers then and farmers now kill off competing wild animals with no remorse. In 2015, our professional hunter in Namibia was granted tags to hunt two leopards on his 25,000-acre conservancy. Meanwhile, the rancher next door, who is allowed to kill any wild animal that threatens his herds, killed seven leopards. Ivan Carter (professional hunter and conservationist) posted the following about African lions on his Facebook page:

> A subject of much recent debate—ban hunting and say goodbye to this species across vast areas of Africa where the only thing protecting them and keeping them there is hunters' dollars—Hundreds of them are poisoned every year by tribespeople.[67]

When humans changed from being hunter-gatherers to farmers, it was a way to improve the protein-flow. When the climate and soil were right, agriculture tended to favor plant production, but harsher environments were suited to meat production. During this "domestication of humans," many people started to believe that it was in some way "better" to eat domesticated plants and animals rather than the wild.

Gradually, farming became specialized, and eventually (with improved transportation and larger urban areas), farming became remote from the cities it supported. People could buy bread ready-made and not have any idea how to grow the wheat. And people could buy meat without having any idea how to kill, dress, and butcher an animal.

Sometimes people think that subsistence is separate from any other type of hunting, but there is an aspect of it in all hunting. Whatever motive for hunting we may have in our hearts, when we eat, we are engaged in the most primal reason for hunting: the nourishment of our bodies.

> Subsistence hunters hunt primarily to stay alive, to live, though they may also hunt for the better quality of game meat over processed meat, and in some cases because of tradition and for the sake of self-reliance.
>
> —Theodore R. Vitali, "But They Can't Shoot Back: What Makes a Chase Fair" in *Hunting—Philosophy for Everyone*

Poaching and Game Ownership

What is poaching, and why do we need to talk about it when we're defining hunting?

In the simplest terms, game ownership creates poaching. Poaching is the illegal hunting, killing, or capturing of wild animals, usually associated with land use rights. (We can even see the highly publicized poaching of rhinos for horn or elephants for ivory as part of an internal struggle for the use of "wildland" and the animals that live on it.) The history of poaching tells us a lot about our current attitudes toward hunting. It helps explain why some people passionately defend the right to hunt.

The word *poach* supposedly has its origins in Old or Middle French or Middle High German, meaning "to trespass for the sake of stealing." It appeared in written form in the early 1600s. Most historians start the history of poaching with the Middle Ages (ca. 1500), but as seen below, game parks (i.e., hunting reserves) go back a lot further, and so, likely, does poaching.

- In Ancient Egypt (about five thousand years ago), on the great estates, there were game parks with animals such as lions, leopards, oryx, ibex, gazelles, baboons, and crocodiles.

- There were luxurious game parks and "paradises" in ancient Persia (about twenty-six hundred years ago).[68]
- The Roman Empire (about two thousand years ago) was well known for hunting for spectacle, but hunting was also considered to be an important part of a youth's upbringing, partly for self-reliance as well as for developing the skills of war.[69]
- During the T'ang period in China (about eleven hundred to fourteen hundred years ago), there were hunting parks that were supposedly accessible to any man if he was a "gentleman."[70]
- In the stories of Marco Polo, there are descriptions of the hunting preserves of Kublai Khan (about seven hundred years ago), which were, by any measure, fabulous.
- Robin Hood (about six hundred fifty years ago) is probably the most famous (and best-liked) poacher in history. Most of us remember hearing stories of him as an outstanding archer who lived in Sherwood Forest. He was reputed to "rob from the rich and give to the poor." But the heart of his dispute with the Sheriff of Nottingham was that he was poaching the king's deer.

In his essay "How Poaching Works," Simon Shadow writes, "As rural poverty was prevalent in the 1700s, many people turned to poaching just to survive. Commoners protected poachers as an act of rebellion, because food was so scarce. Though poaching gangs did provide food to the poor, they were also violent and often greedy, poaching to feed the black market more so than hungry peasants."[71]

Bush Meat and Commerce
From a North American perspective, poaching is strongly disliked by most everyone, hunters and nonhunters alike.

Poaching is no longer seen as a struggle for survival. Nor does anyone recognize it as a type of "peasants' revolt." The poaching we hear about most often is the killing of elephants and rhinoceros for body parts that fetch handsome prices on the black market.[72] However, experts believe this is not the most damaging type of poaching that's going on. The most damaging is commercial meat poaching.

There are some cases where subsistence hunters do poach. Notably (and tragically), they are among the poorest people in the world. Sometimes they are poaching some of the most endangered animals. And sometimes these poor people don't poach just to feed their families directly; they poach to sell the meat. And that's the beginning of a slippery slope.

The U.S. Fish and Wildlife Service (International Affairs) reports that many of "Africa's rare and globally important wildlife species are endangered with extinction. Although habitat loss is a major cause of wildlife decline, the most immediate threat to the future of wildlife in Africa and around the world is the illegal trade and consumption of 'bushmeat.'"[73]

Nearly every modern country faces poaching issues. In North America poachers illegally kill large numbers of deer, elk, black bear, turkey, moose, antelope, cougar, big horn sheep, mountain goat, pheasant, as well as various species of fish, such as walleye, sturgeon, and salmon, and even the ginseng plant.

So, why do hunters need to concern themselves with poaching? There are two reasons:

1. Illegal hunting is part of the problem.

2. Legal hunters are part of the solution.

The challenge for game management is that it's hard to effectively manage the population when it's being reduced by indiscriminate poaching. However, good people are working hard to control the poaching, and good hunters are working hard to support (and even go beyond the requirements of the law) to improve the herd.

The Challenge of Balance

There is no death that is not somebody's food,
no life that is not somebody's death.

—Gary Snyder, *The Practice of the Wild*

As discussed in chapter 1, the idea of balance surely starts with the fact that the predator forms the prey and the prey forms the predator. The grizzly bear can eat many fish, so the salmon need to swim upstream in numbers larger than what can be eaten. The white-tailed deer has amazing hearing and sense of smell, so the black bear needs to stalk stealthily and with the wind in its face.

We remember watching a video in Australia that showed a group of aboriginal women hunting feral cats. The women had no tools, just their bare hands. They stalked the cats. Finally, when they saw a single cat jump into a bush to hide, the women surrounded the bush to prevent its escape. One woman quickly reached in and grabbed the cat by the tail. Before the cat could react, she whacked the animal on the ground to stun it. Then the women threw it on the fire to cook.

Most modern hunters don't have the stealth or the speed of hand to accomplish this. We hunters generally use tools to make the hunt easier and more certain. From the time of our early ancestors, those whose genetics favored successful hunting survived and thrived. For the modern hunter, our ability to improve our hunting tools occurred at a very rapid pace…much faster than genetics can change. So, we have outpaced our prey's ability to respond and keep the playing field even. For example, the "comfort zone" of the white-tailed deer is about one hundred yards.

When we used spears or simple bows, this was a formidable distance. But with the advent of firearms, one hundred yards is an easy shot. The modern hunter still has a challenge in finding a deer, but once it's within one hundred yards, there is no problem killing it quickly and humanely with a rifle.

Now that he can more effectively kill, the hunter also has earned the responsibility to ensure that he does not excessively kill. For example, hunters know that deer become nocturnal when they feel pressured, and game laws do not allow hunting deer at night.

Game management balances the process and supports the survival of the species. So, once again, the prey forms the predator.

Chapter Six—Summary

The "passage from Eden" is a pivotal moment in the history of hunting and humanity. Until that time, there was no other world but the wild and no other means of finding food than foraging and hunting.

"Man the farmer" brought doom upon his hunting brethren. Whether it was physical or cultural extermination, hunting became a nonessential activity, a luxury, and a threat. Agriculture was embraced as the means to feed and control everyone. Hunting became a social activity, usually revolving around elite clubs and hunting grounds savagely protected against commoners who would poach.

Today we have come out of those dark ages, and hunting is again an activity and a lifestyle available to virtually all who would seek it. Now we can share the ancient truths of our ancestors. At the moment of the hunter's kill, we are in tune with our distant past, and everything about life comes into vivid, vital focus. We are here only because our ancestors long ago did what we are doing. Successful predators, we are united across untold eons.

Chapter Seven

Universal Morality and All That Jazz

What Do You Believe?

True hunters see the act of hunting as humbling the human and affirming the continuity of humans with non-human animals.

—Jonathan Parker, "The Camera or the Gun: Hunting through Different Lenses" in *Hunting—Philosophy for Everyone*

Ethics guide behavior within a culture. (Some would say ethics "drive behavior." But it's hunger, sex, and security that *drive* behavior and ethics that *guide*.) There are other things besides ethics that guide our behavior. Among them are cultural traditions, parental standards, and peer dynamics. There are many influences that shape how we behave in each situation.

There is a deep yearning in man to find the one truth that is absolute and universal. We seek the one belief or value or behavior that explains everything and stands absolutely forever. In many cases, religion seeks to fulfill this yearning. But in more secular (less religious) cultures, there are usually some beliefs many people stand by:

- Reciprocity. This is some version of "an eye for an eye" or "do unto others what you would have them do unto you."
- The ethic of outcomes, also known as "the end justifies the means." This says that the consequences of what we do determine whether we've been morally right or wrong.
- The ethic of duty (deontology). This says that an action is moral if it adheres to the rules.
- The ethics of character (sometimes called the ethics of virtue). This holds that we are acting morally if we apply wisdom, fairness, courage, and self-control to our actions. North American First Nations teach these as the "Seven Grandfathers" traditions: bravery, wisdom, truth, respect, love, honesty, and humility.

Hunters often use several (or all) of these ideas to clarify their own sense of what is right. In varmint eradication, it might be "an eye for an eye." In game hunting, a clean kill (fast and as painless as possible) is the "do unto others." In hunting for meat, it might be "the end justifies the means." In following game laws (adhering to the rules), hunters believe they are acting ethically. And most hunters strongly support the ethics of virtue (especially in applying courage and self-control).

The Ethics of Killing
Hunting is killing. One creature dies and is consumed.

Our own bodies are constantly killing to protect us. When your body stops killing the constant barrage of invading microorganisms, you die and decay. To live is to kill, and to kill is to live. In the end, we must get used to it. Anything else can be irrational and dysfunctional to our interaction with the real world.

The fact that humans kill to eat is not news. Many modern humans contract their killing to slaughterhouses. They rarely see meat except in the supermarket. They have separated themselves from the killing process. The hunter takes first-hand responsibility for the meat he eats. The rest of the world likes to start their relationship with a T-bone steak shortly before it hits their plates. The hunter celebrates his connection with the animal from field to fork.

The Ethics of Subsistence Hunting

Subsistence hunting is basic. It is also the easiest form of hunting to defend. If you need the meat to survive, who is going to say you can't have it?

Subsistence hunting also has a long history. It's where we all come from, whether or not we know it or acknowledge it or feel it deep within our souls.

Originally, subsistence hunting had one ethic: conservation of energy. The hunter must get more food value from the prey than the effort he must put into the hunt. Nowadays, nonhunting governments rule most subsistence hunters. Those governments (both local and foreign) try to apply their own values and ethics. Often neither the governments nor the hunters see the other's point of view.

How astonishing it must seem to the Greenlanders…that the English-speaking countries, which only recently desisted from pursuing the great whales to extinction, and then did so only when the slaughter became unprofitable, have awarded themselves the sovereign right to determine acceptable types of hunting for these peoples.

—Peter Matthiessen, "Survival of the Hunter" in *A Hunter's Heart*

The Ethics of Hunting to Kill

There is only hunting to kill. Hunting without intending to kill is not hunting at all.

Photo Safaris

Sometimes people will suggest that hunters should just take pictures and stop killing the quarry. They miss the point.

Taking nature pictures is a lovely hobby and one that many hunters like to do. It's a great reason to walk in the wilderness and enjoy all the plants and animals living there. But it is not a substitute for hunting. In hunting, all participants are completely engaged in the ultimate reality that is life and death. This connection can't be made through photographic hunting.

Fair Chase

"Fair chase" is an idea historians believe US President Theodore Roosevelt (1958–1919) came up with and that the Boone and Crockett Club popularized. It was a stand taken against commercial (market) hunting and intended to attract more hunters to the ideals of conservation. The Boone and Crockett Club has defined *fair chase* as "the ethical, sportsmanlike, and lawful pursuit and taking of any free-ranging wild, native North American big game animal in a manner that does not give the hunter an improper advantage over such animals." They emphasize this is intended to "enhance the hunter's experience of the relationship between predator and prey, which is one of the most fundamental relationships of humans and their environment."[74]

In his essay "But They Can't Shoot Back," philosopher Theodore Vitali[75] comments extensively on fair chase. He says:

> The ethics of fair chase, then, requires these two elements: a careful and scrupulous commitment to conservation ethics grounded in the best biological sciences of the day, and a commitment to foster in

oneself those disciplines that lead to great reflection on—and respect for oneself as an active member of— the life-death continuum.

The Land Ethic

Another ethical issue that is often discussed these days, both by hunters and by conservationists, is the "land ethic." The land ethic is an outline of how humans should care for and interact with the land, the animals, and the plants that reside there. Aldo Leopold[76] (whom we often quote in this book) coined this term.

Our most ancient ancestors likely didn't see land as property. They likely saw it as a place to feed and breed, as did the deer who wandered on it. In our rush to conquer and to own, we focused on making the land belong to us. At the same time, we lost sight of the idea that we belong to the land. But hunters know they have a visceral connection to the land.

Conservation of the natural world runs through hunting mythology. Ancient and modern hunters share this. Hunters—all hunters—are proud of their active membership in the natural world. Ancient hunter-gatherers revered the animal spirits. Neolithic humans celebrated the goddesses and gods of the hunt. Modern hunters fund the current conservation movement. All of these speak to a common morality that underlies our hunting history.

From a hunting perspective, the idea of the land ethic is simply that hunting ethics must be broad enough. They must include not just the animal and its immediate environment but also its entire living community. This is a challenge simply because wildlife managers don't necessarily understand (or control) the larger environment. However, there are good examples of the land ethic being applied to hunting. One is the current work that focuses on understanding the carrying capacity of the land. Another is the study of game movement and conservation of wildlife corridors.

Life Is Messy

Universal morality is a dream-state desire to have everyone—no matter their situation, history, circumstance, or future—subscribe to the same ethics. Whose ethics will we choose?

In practice, ethics are local, cultural, and situational...and changing, ever changing.

Ironically, the more we humans give up on the idea of absolute morals, the more we seem to seek them. As religions (the keepers of absolute morals) give way to more secular lifestyles, we see the emergence of shrill self-righteousness, especially across social media on the internet. This need to be "right at all costs" reflects an underlying insecurity. In *The Righteous Mind*, author Jonathan Haidt reminds us that our moral instincts are inherently judgmental and says that morality "binds and blinds." We seek to be members of a group, and the group morality blinds us to the point of view of nonmembers.[77]

The hunter seeks to identify with like-minded people, and so he enjoys the hunt camp, he reads hunting magazines, he watches the hunting channels on television, he follows Ivan Carter and Jim Shockey on Facebook...and he subscribes to the morality they represent.

> The rules of life are not the same in every place. But of course we want them to be. Our national consciousness drives us to it, and our restless mobility, and the written word, and the electronic net. We love what is uniform: the french fries at McDonald's, universal ethics, and global markets. We shy from the messy specifics of place and from the conflicts that come from living too near the natural world.
>
> —C. L. Rawlins, "I Like to Talk about Animals" in *A Hunter's Heart*

Chapter Seven—Summary

Humankind has always sought to find a "Unified Theory" of "Life, the Universe, and Everything." Explaining, predicting, and justifying all aspects of human existence and behavior remains an academic and philosophical fantasy. For most people, the best answer we can find is within the realm of religious faith, and *that* is certainly beyond the bounds of this study.

Unfortunately, much of our culture today is hellbent on a modern philosophy, a desperate scramble to own, to conquer, and to possess. The mantra of this philosophy seems to revolve around the land "belonging" to us.

Hunters have found deep contentment in the answer that our forebearers understood. Hunters do not "own" the wilderness. When hunters pool vast resources to establish game preserves and protected wilderness parks, they do not intend to "own" that land. The hunters are *in* and *of* the wilderness, inserting themselves into the cycle of life, taking life, and acknowledging (in the deepest and most profound way) that they, too, must pass from this existence.

Hunters have chosen their philosophy of life. They reject the popular belief that the land belongs to us. Those who walk the path of the hunter know that our culture has lost sight of *us* belonging to the land.

Chapter Eight

The Sacred Game

There cannot be two passions more nearly resembling each other than those of hunting and philosophy, whatever disproportion may at first sight appear betwixt them.

—David Hume, *A Treatise of Human Nature*

There have been many spokespersons for hunting. Among them are anthropologists and philosophers, poets, politicians...and hunters. Around the world, there are many types of hunters. The world is large, and there is space for all of them in our embrace. Some of us might not subscribe to each specific cultural practice, but we can admire and enjoy the diversity of this global community.

Henry David Thoreau spent a lot of time working out his personal relationship with the wilderness. He thought a great deal more deeply than most of us would. When some of his friends asked whether they should let their boys hunt, he said, "I have answered yes—remembering that it was one of the best parts of my education—make them hunters."[78]

We've mentioned Teddy Roosevelt before, and he certainly deserves to be noted among our list of great philosophers and influencers in the realm of hunting. He trailblazed the hunting ethic that has become embodied in American game laws. He invented the term *fair chase*. He was instrumental in starting the Boone and Crockett Club, which enshrines the principles of fair chase. His lasting contribution to modern hunting is his ethic of conservation. He said, "In a civilized and cultivated country, wild animals only continue to exist at all when preserved by sportsmen."[79]

David Petersen said in his book *Heartsblood,* "Taken as a whole, [José] Ortega [y Gasset]'s *Meditations* is a compelling call for ethical hunting. Hunting without significant effort, skill, prey compassion, conservation ethic, tenacity, humility, sense of awe in nature, and personal honor...is not hunting."[80]

Aldo Leopold, a contemporary of Ortega, is the author of the popular *A Sand County Almanac*. Many consider him to be the father of wildlife management and of the United States' wilderness system. Aldo Leopold was, among many other things, an outdoor enthusiast. As we discussed earlier, Leopold's main contribution to conservation was the idea of the "land ethic." With due respect to the Indigenous Americans, who had a strong land ethic "before contact" from Europeans, Leopold popularized the idea: "The land ethic simply enlarges the boundaries of the community to include soils, waters, plants, and animals, or collectively: the land."[81]

Where Do You Stand?

When a hunter is in a tree stand with high moral values and with the proper hunting ethics and richer for the experience, that hunter is 20 feet closer to God.

—Fred Bear

Only hunters appreciate that moment of the kill when they are touched deeply in the soul. It can be awesome in the original sense of the word: breathtaking, overwhelming, and amazing.[82]

During the hunting process, prior to the kill, our brain is "balanced." The ancient midbrain and the modern prefrontal brain actively monitor our external world and our inner self. The hunter is aware of himself in the world, and he is aware of the world in himself. A twig that snaps under his foot reminds him that he is a proactive member in the wild. The distant call of the wolf pack reminds him that he is reactive member. When the wolf pack comes closer, a little dump of adrenaline reminds him that he is prey. The prefrontal brain tells him he's a predator today, carrying a gun, and able to protect himself from becoming food.

When the hunter finds his prey and kills it, the whole brain changes. In its purest form, there is no "self" and "other." There is no "past" and "future." There is no "here" and "elsewhere." There is only one "thing." It is the union of the hunter and the world. The ancient brain eclipses the modern brain. And then there is only one brain, and hormones provide its fire.

This is ecstasy. This is unity. This is sacred.

It's Bigger Than Both of Us

Hunting is a small part of the whole picture. When hunters feel that bond to the natural world, they know that they are one small part of a larger thing. One hunter with one animal is symbolic of the relationship between all people and our wonderful world. Hunters are intensely aware of their immediate environment and, at the same time, bonded to the biosphere. Hunters are vividly connected to one animal and, at the same time, merged with all animal conservation. While Aldo Leopold popularized the idea of the "land ethic," it was hunter-gatherers who knew it first. And it was conservationists (mostly hunters) who identified it as a necessary part of modern hunting.

Conservationists provide a voice for animals and their habitat. Most of the funding for conservation comes from hunters. The Boone and Crockett Club is a good example, but there are many more. Among them are Safari Club International, the Dallas Safari Club, Wild Sheep Foundation, and the National Wild Turkey Federation. There are the millions of hunters around the world who make direct donations to conservation and conservation organizations. And those hunters routinely support conservation by buying hunting licenses and game tags. Some hunters also pay large amounts for trophy fees, both fixed price and auctioned. The money goes to habitat protection. It funds game management. And lately, more and more of this money goes toward stopping the poaching of endangered game. Hunters (perhaps above all others) have embraced the need to make poaching stop.

While some eco-elitists cannot see the value of hunting, we can agree on the value of conservation. A powerful argument can be made that we (as a civilization, as a species) should ensure that hunters continue to flourish because without hunters, there would be little funding for conservation.

There can be no doubt that hunting helps local people (especially in Africa) by giving them a way to earn a living. Hunters' trophy fees go toward hiring rangers to protect the game animals. The establishment of conservancies in many areas helps separate the wild game from the domestic ranch animals. And hunters happily support conservation appeals such as "Carter's W.A.R." Zimbabwean Ivan Carter started his Wild Animal Response (W.A.R.) to save his beloved homeland from poachers and wild habitat destruction.

The quiet majority of the peoples of the world are not involved in conservation. The greatest threat to animals is indeed human…the humans who take over the land for housing, industry, and commerce. This is not to say that this is necessarily "wrong," but it is to say that most of the people of the world, the corporations, and the governments are pushing for "progress." Most would rather have another housing development than protect the habitat for animals. Under these conditions, a small minority of the people of the world on the other side of the equation don't stand a chance…and neither do the animals they are trying to protect. And perhaps, in this global equation, hunters are on the side of something noble and good in this dear old world. Something worth fighting for.

Balance

It is worth noting that countries that ban hunting are most likely to have problems maintaining a sustainable population of their animals. It's as though a hunting ban says, "We don't care about our animals."

The most well-known example is Kenya. A beautiful place with wonderful animals, they banned hunting in 1977. The purpose was to save the elephants that were being decimated in the "Ivory Wars." Although there was initially a modest rebound in the numbers, all the gained ground has been lost. Author Glen Martin confirms, "By best estimates, Kenya's wildlife has declined by more than 70 percent" between 1994 and 2014.[83]

In his essay for the *California Magazine*, Glen Martin reflected on the death of Cecil, a thirteen-year-old lion who was killed by a hunter outside the lion's protected area. Martin writes, "Lionizing Cecil makes us feel good, but a trophy hunting ban will accelerate the slaughter." He cites the disaster in Kenya. He quotes Richard Leakey, who said that hunting has never stopped in Kenya, but now it's all unregulated. Leakey says that it's killing, it's snaring, and it's poisoning. It's illegal bushmeat. It's unsustainable.

Glen Martin concludes:

In any evaluation of Africa's wildlife crisis, Namibia must be considered. That's because there isn't a wildlife crisis in Namibia. At the time of its independence from South Africa in 1990, Namibia's game populations were at historic lows, decimated by years of combat between locals and the South African army. The new government wanted to encourage both a wildlife rebound and tourism, but it took a tack directly opposite from Kenya's. Rural populations were organized into communities controlling vast areas of land. Where necessary, the wildlands were restocked with game. Each community was invested with the right to manage its own wildlife resources, subject to certain broad dictates from Namibian national wildlife agencies. In other words, game was commoditized. It could be cropped for commercial meat production; it could be eaten by community members; the rights to hunt trophy specimens of charismatic species could be sold. Suddenly, wildlife had great value for people living in the Namibian bush, and they reacted predictably: they protected their assets.[84]

The foregoing is an exact reflection of coauthors Linda and Keith's experience in Namibia.

The Hunter's Oath

The hunter's ethics can be made clear and distinct with a single statement, a pledge that defines and describes the aim and purpose of the hunting culture. It is the Hunter's Oath. Not all hunters put it in these words, although as a movement, it is advancing. And we hope to give it even more impetus with this book. In any case, most hunters would immediately agree with the ideas embodied by this oath.

The Hunter's Oath
by Capt. Keith A. Cunningham

I pledge on my honor as a hunter to follow the laws of nature, always to hunt ethically, and to obey the laws of the land. I further pledge to develop and maintain the skills required to deliver an effective and humane shot to my quarry. I promise to always do everything within my power to recover my game animal and use it with respect for the life given.

Today, most people in the world do not have the opportunity to be hunters. Chances are, they never will. How those of us who live in the United States gained this privilege is important. It is important because keeping it will depend on how you conduct yourself as a hunter and how you care for the animals you hunt. In short, it depends on your ethical behavior.

—Jim Posewitz, *Beyond Fair Chase*

Chapter Eight—Summary

At the moment the hunter makes his kill, the whole brain changes. There is an existential state in which (in its purest form) there is no "self" nor "other," no "past" nor "future," no "here" nor "there" nor "elsewhere." There is only one "thing" in existence. It is the union of the hunter with the universe. This is the sanctity and ecstasy of the hunt.

At its core, the hunter's ethics revolve around protecting, sustaining, and venerating this profound act. A single statement can make these ethics clear and distinct. It is a pledge that defines and describes the aim and purpose of the hunting culture. It is the Hunter's Oath, in which the hunter pledges to hunt ethically, humanely, and respectfully.

In the next section we'll look at hunting techniques, tools, and skills and tell stories of hunters and how they applied these elements to their own hunts.

Section III

Kudu Eyes

In this section we discuss the underlying skills of the hunter: situational awareness (including "Kudu Eyes"), physical fitness, mental toughness, field craft, and marksmanship) and how these relate both to the hunt and to modern life.

> Our eyes are always pointing at things we are interested in approaching, or investigating, or looking for, or having. We must see, but to see, we must aim, so we are always aiming. Our minds are built on the hunting-and-gathering platforms of our bodies. To hunt is to specify a target, track it, and throw at it. To gather is to specify and to grasp. We fling stones, and spears, and boomerangs. We toss balls through hoops, and hit pucks into nets, and curl carved granite rocks down the ice onto horizontal bull's-eyes. We launch projectiles at targets with bows, guns, rifles, and rockets. We hurl insults, launch plans, and pitch ideas. We succeed when we score a goal or hit a target. We fail, or sin, when we do not (as the word *sin* means to miss the mark).
>
> —Jordan B. Peterson, *12 Rules for Life*

Chapter Nine

How Animals Hunt

As Hugh Brody says, "To make decisions, hunters need knowledge."[85] They gather information about the animals they choose to hunt. They look at where these animals feed and where they water. They look at where the animals bed. They look for their trails. They look for the patterns of other years. And they look for changes.

Hunters compare this year to others. They look at the weather, today's and over the past few days. They try to see a pattern over the months and over the years. They watch for the cycles of animal populations.

> Hunters and gatherers must draw on knowledge that they have accumulated over many years. They must also take careful note of what others say they have seen and done in the last few days. They listen to one another's accounts of hunts and journeys. They watch the sky and feel the wind.
>
> —Hugh Brody, *The Other Side of Eden*

"Empty Hands" Hunting

From bacteria to big game, we are all part of the stuff of hunting. The types of hunting are as varied as the creatures of this earth. "Empty hands" techniques in fighting and martial arts means using only the body without the benefit of any tools. For hunting, it means the same thing. While it's hard for modern humans to picture, our earliest ancestors might have used "empty hands" techniques, just as modern chimpanzees usually do.

Micro-Hunting

If you ever thought that humans invented hunting, you'll be fascinated to see what else goes on in the rest of the animal world. Or, for that matter, what goes on in the plant world. Or, what goes on with the smallest organisms on earth…ones that inhabit our own bodies. Indeed, hunting is truly in the fabric of our being. The cells that we are made of hunt!

In fact, researchers in Europe have created a mathematical model to describe their "spot and stalk" techniques. This enables them to "study not just animal behaviour, but also gain greater insight into the way that killer cells…(large white blood cells patrolling the body) attack colonies of bacteria."[86] And even the smallest of organisms, like bacteria, hunt. Nowadays, microbiology engineers are harnessing this propensity to hunt in order to produce bacteria that hunt down and kill pathogens.

Plants That Hunt

Plants, while they can't give chase, do hunt and eat meat. Often these plants grow in areas where the soil nutrients can't support "vegetarian" plants. The people who study such things have identified several hunting techniques used by these plants:

- Pitfall traps that contain digestive fluids
- Sticky traps where victims are caught in a sticky glue
- Spring traps where rapid leaf closing captures the prey

- Bladder traps that suck the prey into a vacuum chamber
- Funnel traps where inward-pointing hairs force the prey toward a digestive organ

Some of these techniques are also used by animals, including humans.

Early Human Hunters

The human cardiovascular system evolved as part of the physiology of hunters, who regularly ran for their lives.

—Paul Shepard, *The Tender Carnivore and the Sacred Game*

There is strong archeological evidence that our earliest ancestors ate meat before they had tools. Certainly, there is some controversy as to whether they scavenged already dead critters or whether they hunted with "empty hands."

Chimpanzees hunt mostly without tools. As we mentioned earlier, their most common prey is colobus monkeys. And their most common hunting technique is to chase, grab, and disable by biting, disemboweling, or tearing the prey apart. But it's difficult for paleo-archeologists to determine exactly how early humans hunted. The span of time is long. The geographic area is large. The population was sparse. And we've found precious little evidence. Whether they (like modern chimpanzees) occasionally used sharpened sticks to secure their dinner is a matter of speculation. It's not easy to determine when early humans started to hunt by tricking or forcing animals into a pitfall trap or off a cliff. And, since the early tools were organic and perishable, it's not easy to determine when they started using hunting tools like wood and reeds.

Most scientists believe that the significant growth in brain size[87] exhibited by "European Early Modern Humans" is not only attributable to meat-eating but specifically to an improved hunting capability. This improvement is the result of a single technology called "hafting." Hafting is where a sharpened stone, bone, or antler is fastened to a shaft of wood to make a spear. As Valerius Geist writes in *Hunting—Philosophy for Everyone*:

> [The success] of...*Homo sapiens*...was accompanied by a profound improvement in the hunting spear: a sharp stone blade was hafted to the tip of...the stabbing spear...and the thrown spear.[88]

Most early humans were primarily big-game hunters, killing mammoths, cave bears, horses, and reindeer. They hunted with spears, javelins, and spear-throwers. They would have been nomadic or semi-nomadic, following the annual migration of their prey. Although our early ancestors likely used "empty hands" hunting techniques, it's clear that tool use was a significant enabler in hunting meat and, ultimately, in the accomplishments of modern humans.

Tools and Techniques over Time

Sometimes it's easier to understand our own behaviors by looking at what other animals do. When it comes to hunting, there are lots of techniques used by the members of the animal kingdom. Some closely resemble our own hunting techniques.

Chemicals

The bacteria we mentioned earlier work in large groups. Their behavior is reminiscent of urban gangs, but they finish the job with chemical warfare. As Marlene Cimons reported for the National Science Foundation:

> When food is plentiful, these bacteria move in coordinated swarms, called ripples, often containing thousands of cells...[These] secrete enzymes into the environment to kill their prey...[They] digest it

outside their structure before taking in the resulting nutrients...

Single cells can't produce enough of these antibiotics or enzymes to effectively kill their prey, which is why they hunt together as a group.[89]

In fact, many organisms use some type of chemical to kill or at least to immobilize their prey, among them Huntsman spiders and venomous snakes. Early humans used poison on the tips of their spears or arrows, and some tribal hunter-gatherers still do. They also hunt in groups, enhancing this technique. As Gateway Africa, a safari and conservation website, describes:

Depending on the size of an animal it can take from a few minutes for a hare to up to a few days for a big animal like a giraffe. But usually when bigger game are shot, the hunters try to get as [many arrows as possible] in the animal and try also to deliver the arrow to a jugular or main artery. After shooting the animal, the San [African Bushmen] will wait for a few minutes/hours, waiting for the poison to take effect before tracking of the animal will commence which can last for days and over many kilometres.[90]

Netting

Another animal hunting technique is netting. Netting can be passive (set and left for animals to encounter) or active (launched when the target animal is within reach). For example, the so-called net-casting spiders spin a rectangular net instead of spinning webs to catch their prey. They hold this net in their front legs, waiting for a prey animal to come along. When given the opportunity, they push the net over the prey, capturing it. (Then they bite the prey to paralyze it before they consume it.)

Human hunters use netting for small game, birds, and fish. Probably the earliest "recorded" use of nets for catching small game is from Europe thousands of years ago. Imprints made on clay indicate that netting was used by the Gravettian peoples (who populated areas from southern Spain to southern Russia during this time).[91] Nets are an efficient way to catch large numbers of small animals, and these likely presented a more consistent food supply than big game. This regular source of protein may be one of the reasons why these people were able to live in larger, more settled communities.

Baiting and Trapping

Baiting is widespread in the animal kingdom. There is a type of fungi that captures and digests nematodes (small, worm-like organisms) by trapping them. These nematophagous fungi include the fast-growing shaggy mane, popular among mushroom gatherers for its highly distinctive form.

One particularly vivid example of a bug that survives using baiting to obtain food is the assassin bug. The assassin bug cleans up garden pests by baiting, poisoning, and eating them. The bug attracts its prey with a sugary substance and poisons the prey with a toxin that paralyzes the victim, then liquefies its innards. Finally, the assassin bug sucks up its "protein drink."

The angler fish has a lure that protrudes from its face, drawing prey in so that it can catch it in its large mouth. The alligator snapping turtle has a natural lure, which National Geographic describes this way: "Its tongue sports a bright-red, worm-shaped piece of flesh that, when displayed by a motionless turtle on a river bottom, draws curious fish or frogs close enough to be snatched."[92] Some snakes use the tips of their tails similarly. Snakes generally lure frogs and lizards, although the luring of birds has been reported and luring of moles and shrews is suspected.

Early man likely used field-expedient baiting. When he found another animal's kill site, he might have scavenged. Or he might have leveraged his opportunities by killing other animals attracted to the site. The ancient water hole was also a likely-used bait. Both modern humans and animals use the water hole as a popular place to find prey. It's easy to imagine that as soon as man learned to store food, there would have been animals trying to steal the stores…a form of passive (or unintentional) baiting.

Modern hunters use baits commonly in fishing, sometimes to attract fish to a feeding site (chumming) and sometimes to lure the fish to a hook. Bird hunters use decoys to attract waterfowl and turkeys. Many modern deer hunters use food plots (whether purposely planted or as part of a farm) or scents to attract the animals to a location under their observation. Bear hunters often use food as bait.

Many people use baiting in conjunction with various forms of trapping, especially with

- cage traps (for live capture),
- spring traps (like mouse traps),
- deadfall traps (often used by Native Americans), and
- body-hold traps (used by professional trappers for fur-bearing game).

Other forms of traps (like snares and pits) usually don't use baits but instead rely on either the exploitation of game trails or driving the quarry. Another type of trap is the glue trap, sometimes baited and sometimes placed in an area of opportunity (like the sticky traps for cockroach or mouse control).

Spot and Stalk

Earlier, we told the story of the Indigenous Australian cat hunters who used the "spot and stalk" technique when hunting feral cats. This technique is often used in the animal kingdom. Hawks fly circuits looking to spot their prey and then swoop down to capture their meal. Many seabirds do the same thing. For land mammals, the "spot" part of the hunt might be more aptly described as "smell." Many members of the dog family stalk their prey based on the smell of the track. In fact, we have "trained" our domestic dogs to track game (or lost people). This is an adaptation of the natural hunting behavior of the dog.

> My dog, by the way, thinks I have much to learn about partridges, and, being a professional naturalist, I agree. He persists in tutoring me, with the calm patience of a professor of logic, in the art of drawing deductions from an educated nose. I delight in seeing him deduce a conclusion, in the form of a point, from data that are obvious to him, but speculative to my unaided eye. Perhaps he hopes his dull pupil will one day learn to smell.
>
> —Aldo Leopold, *A Sand County Almanac*

Among human hunters, the "spot and stalk" technique is one of the most popular modern hunting methods. Hunters in Africa use it for plains game and dangerous game. Hunters in North America use it for deer and sheep. Hunters around the world use it to enhance the pleasure of the hunt. Most hunters enjoy spending time in the wild, and looking for game gives purpose to the hike. Once the game is spotted, the stalk is on!

The pleasure we humans find in stalking is deeply rooted in our psyche. As Mary Zeiss Stange writes in *Woman the Hunter*:

> [A professional hunting guide] had remarked an interesting fact of his experience: that, when stalking an animal, people visibly changed, their pace becoming more stealthy and deliberate, their breathing and facial expressions altered. Asked if this was apparent only in seasoned hunters, he replied that it was a

constant in all his hunters, beginners and old pros
alike. He added that it was interesting that women
and children who had never hunted before had all
the instinctive movements that many people think
they learn only through years of experience.[93]

Ambush

Ambush is a hunting technique often used in the animal kingdom and usually associated with pouncing. Once the predator has
pounced upon the prey, the predator still needs to immobilize and
consume, and the techniques for this vary widely. Biologists usually characterize carnivorous plants as ambush predators.

Among insects, the praying mantis is an ambush predator
that is well-known to most of us. It stays very still until an insect
comes near and then uses lightning-fast reflexes to grab the prey.

Many fish (like pike, groupers, and frogfish) use ambush
and usually suck their meals into their mouths. Reptiles like crocodiles are well-known for lying in ambush and pouncing on their
prey. There are also lizards, turtles, and snakes that use this technique. National Geographic describes the hunting techniques of
Komodo dragons as ambush predators: "Komodo dragons rely on
camouflage and patience, lying in wait for passing prey. When a
victim ambles by, the dragon springs, using its sharp claws and
serrated, shark-like teeth to eviscerate its prey."[94]

Most members of the cat family pounce on their prey. You
may have witnessed this behavior in the domestic cat, particularly
if you have entertained your pet with a laser pointer. We have spent
many hours watching foxes pounce on mice and voles. The polar
bear sits quietly by the seal's breathing hole in the ice and waits,
often for hours, until a seal's nose breaks the surface of the water.

The modern hunter will monitor a bait or sometimes a game
trail (where the game animals move as they go from bedding to

feeding to watering areas). The hunter uses either a ground stand or a tree stand, or he simply camouflages himself amongst the rocks or bushes. Many hunters prefer this style of hunting as it affords a lovely opportunity to sit and enjoy the wilderness in peace. They are pleased to leave the rest of the world behind even if only for a few precious hours.

> In many hunting-gathering peoples the trial includes a time alone. Wilderness solitude is not a test against the elements, discomfort, and deprivation, for, after all, the young hunter is already an expert in survival. His purpose is more lofty…It is to receive the full weight of the cosmos on his head without the shield of society. It is perhaps difficult, in surroundings constructed by men, to imagine such an encounter except as the threat of physical danger, but the personal significance of this experience in a contemporary form is shown in some measure by the Outward Bound program, [where after three days alone in the wilderness]…its many enthusiastic graduates do not say, "It was great—I survived!" They say, "It was great—I got to know myself."
>
> —Paul Shepard, *The Tender Carnivore and The Sacred Game*

Camouflage and Mimicry

Camouflage is an essential part of stalking game. The predators of the world are more successful if they can "hide in plain sight." The predator must blend in with the environment visually as well as conceal any odor or sound that might alert its prey. (And of course, the prey also use camouflage to hide from the predators.)

But if you can't hide yourself completely, another alternative might be to look like something you're not. Both predators and prey use mimicry to hoodwink their opponent. Here are some examples of camouflage and mimicry:

- The chameleon is a camouflage expert, changing color to match its background so it is very difficult to spot.
- Many birds and animals use molting to change colors for the season, such as the ptarmigan, the snowshoe hare, and the arctic fox.
- Many animals use color and pattern to camouflage themselves in their natural habitat, such as the whip-poor-will, which not only looks like a pile of gravel on a road but also uses the darkness of night.
- Many fish are dark colored on top and light colored on the bottom. Some predators (birds) looking down on them find it hard to see them with the depths of water below. Other predators (other sea animals) looking up at them find it hard to see them against the water brightened by the sky.
- Some animals minimize the shadow they cast by flattening themselves (horned lizards) or by orienting toward the sun (butterflies).
- Many predators use camouflage to help them approach prey, such as leopards who have a disruptive pattern or tigers whose stripes help them move unnoticed through the grasses.

Human hunters often wear camouflage clothing. This is not new. Both ancient and contemporary hunter-gatherers often wore animal pelts and bird feathers, as well as face paint. Some modern hunters make their own camouflage from old military clothing trimmed with hessian (strips of burlap). Most prefer modern specialty clothing ("3-D suits," ultra-realistic photo-cam, or military disruptive patterns). This clothing often comes with an ultra-quiet exterior finish or odor-eliminating lining.

And if you thought that humans were the only ones to *wear* their camouflage, you might want to look at the inventiveness of other animals. Some decorate themselves with materials such as twigs, sand, or pieces of shell from their environment. This can break up their outlines, conceal the features of their bodies, and match their backgrounds. For example, a caddis fly larva builds a decorated case and lives almost entirely inside it. A decorator crab covers its back with seaweed, sponges, and stones. The nymph of the predatory masked bug uses its hind legs and a "tarsal fan" to decorate its body with sand or dust.

> Hunters have been impersonating their prey forever. And not only human hunters: I've read credible accounts of both European brown bears (grizzlies) and Siberian tigers consciously imitating the "roars" of red deer (Eurasian elk) in order to lure rut-crazed stags to ambush. This is utterly natural.
>
> —David Petersen, *Heartsblood*

Another type of camouflage is staying still. Movement is one of the most eye-catching reasons things are seen. When animals are hiding, they use no movement or only stealthy movement, or they move to mimic their background or other animals. For example, deer fawns can lie very still and stay hidden, aided by their spotted coats (and if they are very young, by their minimal scent). The large cats are well-known for their stealth, moving slowly and quietly while they close the sprinting distance to their prey. The leafy sea dragon, a relative of the seahorse, has lobes of skin that give it the appearance of a piece of seaweed as it floats through the water. It can also change color to blend in.

Countershading is a type of camouflage that provides a bit of optical illusion. The parts of the body or face that are normally in the foreground are darkened, and the parts that are usually in shade are lightened. This can help the animal look flatter and softens the edge of its body as it relates to the background. The bow hunter makes good use of face paint to camouflage the human face, which is normally all too recognizable as a predator. Our hunting ancestors may well have used ocher and charcoal to accomplish the same thing.

In addition to using camouflage, modern hunters often use game calls to mimic the sounds and calls of their quarry. The deer hunter may use a "fawn in distress" or a "doe in heat" call and may also rattle antlers to mimic the sound of bucks fighting. The waterfowl hunter and the turkey hunter often use calls to mimic (and attract) their quarry. The predator hunter usually uses calls of the prey animal; for example, the coyote hunter will use the sound of a dying rabbit to attract the coyote to what he thinks will be a free lunch.

Margays (nocturnal cats in Central and South America) prey on the tamarin monkey. The cat mimics the sound of a distressed baby tamarin. When an adult tamarin comes to investigate, the margay attacks and eats the adult monkey.

One of the most interesting uses of sound mimicry in the animal kingdom is that used by the drongo. This African bird feigns the warning calls of meerkats, causing them to flee and to abandon their meal. Then the drongo steals their food. It would be comparable to a person yelling, "Fire!" in a grocery store and then filling a cart after the shoppers and employees ran out the door.

Chasseur
The chase is a widespread technique in hunting.

Although we usually think of land mammals when we think of a chase, many other animals successfully use some form of chase. In the insect world, the tiger beetle is known as the fastest animal in the world. It can run at only about 5 miles per hour (mph) or 8 kilometers per hour (kph), but proportionately, that's the equivalent of a human running over 300 mph (480 kph). The beetle's speed is faster than its eyes can process visual information, so you could say it's "running blind." Indeed, it must periodically pause to let its eyes catch up so it can relocate its prey. Birds (such as owls, falcons, eagles, and hawks) are also known to chase their prey. However, they usually surprise them so that the chase is short-lived. Sea lions are known to chase fish at speeds of up to 25 mph (40 kph).

There are a couple of types of chase. They are like the sprint and the marathon. The sprint is all about speed, and the marathon is all about endurance. The cheetah is best-known as the top sprinter in the animal world, commonly running at over 50 mph (80 kph) with bursts up to 75 mph (120 kph). The cheetah is also famous for its acceleration: 0 to 60 mph (100 kph) in three seconds.

An article in *Popular Mechanics* describes the top marathoners in the animal kingdom. Among them are horses, humans, dogs, camels, Pronghorn antelope, and the ostrich. To illustrate:

> Each year, during the Iditarod, packs of 12 sleigh dogs travel more than 1000 miles [1600 kilometers] in 15 days. The animals working together pull the sled at speeds around 15 mph [24 kph] for up to six hours at a time.[95]

If running a marathon, these huskies would cross the finish line ahead of human competitors.

Grizzly bears are well-known predators. They use both ambush and chase techniques. They commonly use ambush for salmon during the spring run when the fish are swimming upstream. They also ambush mature bull elk when the elk are distracted by the rut. But grizzlies also use "the chase" to run down mature deer and elk in the early spring while there is still snow. Cougars similarly kill deer and elk in winter by running them down in the snow. In fact, white-tailed deer in snowy seasons will congregate in winter yarding areas, where the snow is packed and the herd can defend against hungry wolves.

Predators may use the chase technique simply to pursue the game until they catch it with claws and fangs. The chase may also be part of another hunting tactic, such as pushing the prey to a specific location. As we've mentioned, some Native American Indian tribes were known for the buffalo hunts that pushed the herd over a cliff. In other cases, the chase would push the prey to a choke point (a narrow pass or a waterway) in order to ambush it. And sometimes, the objective of the chase was to produce a stampede, which would cause the herd to abandon the slower members, making them easy prey.

Animals that give chase sometimes work singly and sometimes in groups. Most cats work alone, but female lions often hunt as a group and sometimes use a relay form of chasing. Dogs also hunt in packs, and wolves are well-known for taking much larger prey as a group than they ever could as individuals.

Another type of chase where dogs are commonly used is called *battue*, where the dogs "beat the bushes" to flush the deer. In England, hunters use "beaters" to push birds from cover, while in America, hunters commonly use bird dogs. In fact, there are many forms of hunting where humans use the natural hunting abilities of dogs to point or to flush, herd, drive, track, pursue, or retrieve prey. It's no accident that ancient dogs became ancient man's best friend. In modern hunting, hunters still often use dogs, but where they are not available (or not legal to use), sometimes people (called "doggers") will "drive" the game to push it past riflemen who are already in position.

As we've mentioned earlier, persistence hunting is the type of chase where the predator tracks and pursues the prey to exhaustion. Our very earliest ancestors likely used this hunting technique. The Bushmen of the Kalahari still hunt using this method.

Most modern humans don't have the fitness required to hunt by persistence, but we still have the physiology. As Paul Shepard points out in *The Tender Carnivore and the Sacred Game*, most cats hunt by stealth and high-speed charges, while dogs hunt by relentless pursuit over long distances. Man's "circulatory and endocrinal systems are like those of dogs," suggesting our hunting developed "by endurance and strategy rather than fleetness and surprise."[96]

> We're designed for persistence hunting, which is a mix of running and walking. What's built into that kind of running is a sense of pleasure. You are designed and built and perfect for this activity, and it should be enjoyable and fun.
>
> —Christopher McDougall, *Born to Run*

Tools

The original tools of man and animal alike were their five senses plus hoof and horn or claw and fang. Our ancient ancestors may well have had better senses (particularly vision and smell) than we do now, but there is no doubt that *Homo sapiens* would not have fared so well against his prey without also using his wit and his tools. Even the chimpanzee uses a tool as he "fishes" for termites with a stick or straw. The chimp uses a sharpened stick to stab bush babies. However, he also uses claw and fang when he snatches small game.

> You are the weapon; everything else is just a tool.
>
> —Anonymous, as quoted in *On Combat*

In *The Tender Carnivore and the Sacred Game*, Paul Shepard eloquently describes the deep roots of toolmaking in our ancient ancestors.

> Man did not discover the art of toolmaking—he
> inherited it…The most successful tool ever made…
> was the hand-ax.[97]

The hand-ax was not hafted (tied to a handle) at first. It is ancient, and it's still used today. Shepard points out that the neural feedback systems of hand and eye with the use of tools is more primate than human. He compares man the hunter using a piece of flint to dissect his prey to the lion or wolf that uses muscles and fang.[98] As we've said before in this book, our idea of ancient man's tools is based on what we've found at dig sites. We find only those tools that survived the millennia. The stone tools we have found take a significant amount of skill to make:

> There emerged an exquisite body of work of a very
> high caliber, knapping (chipping) techniques so
> refined that it takes many weeks of practice and
> instruction for a modern man to master them, and,
> indeed, some of the techniques have been lost.[99]

Shepard also discusses fire as a tool and says the first hearths may date back to our most distant past. And he suggests that recent archeological discoveries (like the lunar notations found on Paleolithic materials in Spain) may indicate a much richer pre-literate intelligence than we've so far acknowledged. He writes, "Our eyes would be opened a little better if we thought that literate meant 'leaving written record' instead of 'educated.'"[100]

> Tool use is certainly no human invention. Analysis of motor patterns indicates that throwing, clubbing, and incipient weapon use are extensions or derivatives of the general intimidation displays typical of primates. Among chimpanzees, as among men, this skill is improved by experience and practice.
>
> —Paul Shepard, *The Tender Carnivore and the Sacred Game*

Projectiles

The early hunting tools like the stick and the spear, the rock and the ax, not only provided a force-multiplier for the strike, but they also provided the hunter with a safety distance from the prey. Throwing tools, while adding to the distance, usually reduced the force and the accuracy of the strike and would have required additional skills.

Among animals, there are many examples of projectile use, both for hunting and for defense:

- There are worms, spiders, beetles, ants, and termites that spray various noxious fluids on their prey.
- The archerfish uses a jet of water to reach land-based insects and other small animals. The fish uses his "water gun" (specialized mouth parts) to shoot down his prey with a jet of water. When the quarry lands in the water, the archerfish is waiting to enjoy its meal.

- Most of the vertebrates that use projectile fluids (from geckos to camels) use them in defense. However, there are snakes that can project their venom as part of predation. While they're not always considered projectiles (although some biologists call them "tethered projectiles"), frogs are well-known for "shooting" their tongues out to capture prey. They also use "glue" as part of the trap.

- Some tarantulas have a dense covering of hairs that they sometimes use as protection against enemies. They can kick these hairs off and then flick them at a target using their back legs. These fine hairs are barbed and designed to irritate and can be lethal to small animals such as rodents.

- The antlion digs a sloping pit in the sand and then lies in wait at the bottom of it. Small prey slip into the pit. If the prey tries to crawl out, the antlion throws sand at it, sending it back down into the pit.

- The California ground squirrel kicks sand into predators' eyes. This doesn't kill, but it distracts predators, such as snakes, so the squirrel can make its escape.

- An African elephant has been observed using its trunk to throw various materials (mud, soil, vegetation) at an interfering white rhino.

- Some primates (including humans, bonobos, chimpanzees, gorillas, orangutans, capuchins, certain gibbons, and perhaps some baboons and Japanese macaques) are known to throw projectiles (such as sticks, rocks, and feces). In a BBC news story, a male chimpanzee in a Swedish zoo was seen stockpiling stones, which he later used to throw at visitors in a planned attack.[101]

- The aptly named pistol shrimp (also known as a snapping shrimp) hunts by using an air-bubble as a projectile. The pistol shrimp is altogether an amazing little animal. It lies in wait, and when it senses passing prey, it releases a "shot." The shot is a cavitation bubble that, when it collapses, produces a loud sound, and the heat it generates produces a flash of light. (The shot can reach speeds over sixty miles per hour, and the "snap" is over two hundred decibels, which is louder than a gunshot.) The shot stuns or kills the prey, which the shrimp then feeds on.

Projectiles are part of the long history of human tool development. We can probably describe the history of hunting arms as a search for something that provides a quick, targeted kill while minimizing danger to the hunter.

Ballistics (in terms of leveraged projectiles) may have begun with the sling or the atlatl. The sling is an ancient weapon known to be used in the Stone Age but is likely much older. The atlatl (used to throw darts or spears) gives the user a mechanical advantage and is estimated to launch projectiles at speeds over ninety miles per hour.

Another such tool is the bow, which dates back to our distant past. But however long ago the invention of the bow may be, spears or darts are much older. It has been suggested that some early points from Africa might have been arrowheads and that use of bows and arrows may date back over seventy thousand years ago.[102]

The timelines for the human use of tools and tool improvements change regularly. As archeologists find new artifacts, our understanding of our hominid ancestors is constantly challenged and expanded.

Over time, two innovations worked together to produce modern firearms: improved methods of launching a projectile and the ability to work metal. As blowguns yielded to air guns and stone axes yielded to edged metal weapons, the use of gunpowder overtook hunting technology. Nowadays, hunters can choose from a wide array of hunting arms, both traditional and modern. There are people who still use spears and knives, bows of all types (from longbow to modern crossbow), and firearms from muzzleloaders to highly tuned modern rifles. All these technological advances have worked toward improving lethality, especially accuracy and terminal energy at a distance.

One excellent consequence of modern hunting arms is that people who would not have been able to hunt with traditional arms (because they favor able-bodied people of a specific size or strength) are able to go confidently to the field. This has opened hunting pleasures to more women and to people of both genders who have physical limitations.

We Kill to Be Humane

Modern hunting techniques (used by humans) strive to both immobilize and kill by a single means. The rifle shot is intended to kill as is the hunting arrow, and both are intended to immobilize within seconds. Other animals have no such delicacy; indeed, it is only a recent refinement of modern hunters. For the most part, other critters incapacitate and eat, and killing is incidental. We kill to be humane.

Immobilization

For nonhuman animals, there are many ways they immobilize their prey:

- Siafu ants crawl into their prey's lungs to feed, which results in death by suffocation.
- Snakes and other reptiles, as well as many insects, use poison.
- Anacondas use squeezing, suffocating, and drowning.

- Crocodiles use dismembering and drowning.
- Electric eels shock their prey with voltage, sometimes just stunning them and sometimes killing them by electrocution.
- Some birds drop shellfish on rocks, primarily to open them but consequently killing them.
- The large cats catch their prey by chasing and slow them down with as little as a single claw; they then use suffocation or strangulation to immobilize. Although many references state that most big cats kill before they start to feed (and certainly they are well equipped to do so), we have all seen the footage of the lions eating a Cape buffalo that is still very much alive. (Notably, the jaguar bites through the brain to immobilize.)
- Many primates use mauling, which immobilizes and, while it doesn't kill immediately, can certainly send the prey into shock, and death follows.

A vivid example of the use of immobilization in hunting is the mother teaching her youngsters to hunt and kill. In many predator species, the parent immobilizes the prey to some degree and brings the captured animal to the youngsters, where the parent then encourages the juveniles to "play" with it. If the prey tries to run away beyond the youngsters, the mother will recapture it and return it to the "classroom." Eventually, the youngsters will successfully immobilize it and have their reward.[103]

Cute Killers

Sometimes people have "forgotten" that some of the cutest animals are, indeed, killers. After decades of seeing them as stuffed animal toys or cartoon characters, humans no longer consider as fearsome some of the deadliest predators. These animals include bears, cats, wolves, and monkeys.

Bears

Bears are often criticized for just being the animals they are. A number of highly publicized bear killings were a result of the prey (hikers, campers, observers, and self-proclaimed naturalists) being somewhat naive about the true nature of a predator.[104] Just as anthropologists have noted that omnivores usually eat more meat when it's not berry season, so we could possibly predict that bear predation is likely to occur in cycles. And it does. As the old saying goes, bears eat when they're hungry.

While it is likely true that territorial and defensive bear behavior may be relatively hard to predict, it is possible that there may be more bear kills than necessary simply because of the arrogance of humans who don't believe bears are predators. It may have all started with the story about *Goldilocks and the Three Bears,* which certainly made the bears seem just like us, not predators but a loving, porridge-eating family living in the woods. The teddy bear, an adorable stuffed toy with neither fangs nor claws, accelerated the transformation from wild animal to cute pet. Published in the 1920s, the stories of Winnie the Pooh and his animal friends likely also played a role in the domestication of the wild bear. The gummy bear candy (originating in 1920s Germany) lets us turn the tables and eat the bears! And in the 1980s, Ewoks—cute and friendly little bear-like critters—appeared in the *Star Wars* world.

Bryan Nelson describes bears as follows:

> Bears are some of the most lovable large carnivores
> in the world, often the subject of childhood tales and
> treasured as teddy bears. It's a strange association,
> given that they are also on the shortlist of animals
> known to hunt and kill humans. Grizzlies and polar
> bears are the most feared, but all large species of bear
> can potentially be dangerous—even the vegetarian
> giant panda.

A study in the *Journal of Wildlife Management*...found that solo, hungry males—not mothers with babies—are most often the ones that kill.[105]

Bears are good killers. They may be opportunistic (although there are certainly reports of bears stalking humans), but once a bear has its fangs and claws into its prey, it is hard to change its mind...and hard to survive the attack.

Cats

Other adorable little animals that can kill you are numerous, but the cat family is probably the cutest. The big cats are very much underestimated by everyone except those who live among them in the wild. Big cats (lions, tigers, leopards, jaguars, cheetahs, panthers, pumas) are normally carnivorous predators. However, many people believe that they only kill when they must and only kill the weak and sick and old. In fact, the big cats kill to eat—and they may hunt anything that looks like a good (or easy) meal.

Lions (like many cats) are scavengers, reportedly deriving about half of their diet from carrion. The Disney movie *The Lion King* aside, *Walker's Mammals of the World* is the somewhat more authoritative source. *Walker's* indicates that "most dead prey on which both hyenas and lions feed are killed by the hyenas rather than the lions."[106]

But lions are also ferocious killers, successful with dangerous game (such as crocodiles and Cape buffalo), plains game (such as wildebeest, zebras, kudus, oryx, eland, and giraffes), livestock (such as cattle), and a variety of people (ranchers, refugees, villagers, tourists, and guides).[107] Oh yes, and lions kill other predators such as leopards, hyenas, and wild dogs, although they seldom eat them after killing them. Lions tend to dominate the smaller felines such as cheetahs and leopards when they hunt the same areas, stealing their kills and killing their cubs and even killing the adults when given the chance.

Tigers are also marvelous killers, and although they hunt alone, they take some very large prey. Their usual prey are deer, boar, and buffalo, but they have also been known to prey on other predators (such as leopards and crocodiles) as well as Asian elephants and Indian rhinoceros. Siberian tigers take on black and brown bears. When in close proximity to humans, tigers will also sometimes prey on such domestic livestock as cattle, horses, and donkeys. And they can become man-eaters. In fact, as Peter Capstick says in his book *Death in the Silent Places*, the impact of this animal over the past four hundred years of closely recorded history is almost unimaginable.

The best thinking on the matter, distilled from expert sources, indicates that tigers alone have eaten at least five hundred thousand Indians and certainly a minimum of one million people over the entire Asian continent in the last four centuries.[108] As we mentioned in the introduction to this book, Bruce Siddle's research on predation shows that, according to British records, in India from 1900 to 1910, a single decade, over one hundred thousand people were killed by tigers.[109]

Hunters often quip among themselves that leopards "deliver more stitches per second" than any other animal. Leopards are versatile, opportunistic hunters and have a broad diet. They feed on a greater diversity of prey than other members of the genus *Panthera* and eat anything from dung beetles to common elands. Their diet consists mostly of ungulates like antelope, gazelles, and impalas, but they also feed on primates (mostly monkeys but also chimpanzees and gorillas). They enjoy rodents, hares, birds, lizards, and fish. They stalk their prey silently, pounce on it at the last minute, and strangle its throat with a quick bite. They may eat small prey right away, but they'll often cache larger prey in a tree.

Like other big cats, they also kill (but usually do not eat) competitors, especially the cubs of other cats, as well as hyenas and wild dogs. Leopards prey on livestock, and many are annually killed by ranchers protecting their cattle. Occasionally leopards become man-eaters as well.

Wolves and Coyotes

The human love affair with dogs makes the hunting techniques used by wolves and coyotes of particular interest.

A technique reportedly used by coyotes (who usually hunt alone or in mated pairs) is to "kill by wounding: by sinking a tooth in an animal, then following it for a week as the wound festers to gangrene. The coyotes trailing it, staying close, before finally moving in."[110] Wolves, however, hunt cooperatively in packs, so capturing and quickly immobilizing their prey is the objective. Wolves may shake smaller prey to break their spine. The pack works cooperatively with larger prey. They start eating as soon as they bring the victim down.

> Indeed, it is not uncommon for a large animal to be still alive when the wolf pack or hyena clan is already munching on its intestines. Of course, death follows soon after, due to shock or blood loss, but still, the idea of being alive while a snarling group of voracious predators feeds on your entrails is particularly disturbing to most people.[111]

Many farmers, ranchers, and hunters believe that some animals, especially wolves and bears, kill for sport. This makes sense from a territorial perspective, and it makes sense for animals to kill when the opportunity presents itself and come then back later to feed.

In the same sense that animals get pleasure from sex and security, it seems logical for the predator to get pleasure from the hunt. It would be like the kind of enjoyment we get from a cozy fire and a fresh-baked cookie. We get a wave of pleasure from somewhere deep inside, a wave that says, "All is right with my world." This is the hormone-based pleasure that comes from the endocrine system and that our modern brain (the prefrontal part of the brain particularly) spends many hours explaining and justifying.

Wolves are also known to kill domestic animals and to such an extent that many governments have programs to compensate farmers who suffer livestock predation. Sheep are commonly killed, often *en masse*. Coyotes are known to pester a cow that's trying to deliver a calf and will start eating the calf before it has completely emerged from the mother. Coyotes attack other animals around the farm, including the pet dog and cat, and they do attack humans.[112] Urban coyotes are losing their fear of humans, which is further worsened by people intentionally or unintentionally feeding coyotes. In such situations, some coyotes have begun to act aggressively toward humans, chasing joggers and bicyclists, confronting people walking their dogs, and stalking small children.

Wolves are a much greater concern because they are big enough to kill a human adult and, since they hunt in packs, could even threaten groups of humans. However, because wolves are less social than coyotes, they tend to avoid human contact.[113] (Having said that, the authors always exercise great caution when we're hunting and especially when we're field dressing our deer.)

So why do North Americans often think wolves aren't dangerous? Here's a little history:

- "The origin of the 'harmless wolf' myth can be traced to a highly respected Canadian biologist, Dr. Doug Clark. He investigated the killing of people by wolves in Europe and concluded in an unpublished paper, 'The Beast of Gèvaudan,' that while such attacks were real, rabid wolves caused them all."[114] (This was wrong.)

- "When Farley Mowat published his 1963 book *Never Cry Wolf*, it was heralded by environmentalists from his native Canada all the way to the Soviet Union. His real-life account of wolf behavior in Canada seemed to shed new light on their prey, their behavior and their role in an ecosystem." [115] There were many disclaimants, including the Inuit (who said the wolves weren't accurately portrayed), and in a 2012 interview, Mowat acknowledged that the work was fictional.

- By the 1970s, the fear of wolves was largely counteracted by the emergence of a "pro-wolf lobby aiming to change public attitudes toward wolves, with the mantra 'there has never been a documented case of a healthy wild wolf attacking a human in North America.'"[116] (Also wrong.)

Following a wolf attack incident in Icy Bay, Alaska, biologist Mark E. McNay "compiled a case history that describes 80 wolf-human encounters in which wolves showed little fear of people...Thirty-nine cases showed elements of aggression among healthy wolves, 12 cases involved known or suspected rabid wolves, and 29 cases document fearless behavior among non-aggressive wolves."[117]

There is also an infrequently reported fatal attack that took place near the authors' home area in 1996.[118] Trisha Wyman had just started a new job at the Wolf Centre preserve. In her first week, fellow employees discovered Wyman's body in the wolf enclosure. Her clothing had been totally removed, and strips of it were strewn around her body. The body itself was relatively intact, though with multiple bite wounds and some flesh missing from her extremities. Authorities later destroyed the wolves in order to examine them for rabies. The results were negative.

The authors of this book have all witnessed the reaction of animals to nearby wolves and have seen the results of wolf kills in the woods. The prey is usually completely eaten but not always. Coauthor Keith once came upon a deer kill that looked a lot like the Trisha Wyman kill site. The area was thoroughly trampled, and deer's hair (in bite-size clips) lay all around the body. The deer had not been eaten and didn't appear to have any serious injuries. But it was dead.

Was the killing behavior driven by blood lust? Did the wolves kill the deer "for fun"? Again, "for fun" is hard for us to judge. But just as some predators have the instinct to pursue anything that runs from them, we believe that some predators have the instinct to kill anything that looks like a good (or easy) meal even if they're not hungry. And just as our hormones reward us for good survival behavior like sex, we believe our hormones reward us for the other good survival behaviors like hunting.

Chimpanzees and Killing

Monkeys and apes are pretty cute…kind of a furry version of ourselves, right? We humans are often compared to the animal most similar to ourselves in nature, the chimpanzee. And how do they kill?

The Jane Goodall Institute of Canada website describes one kill as follows: "The [chimpanzee] hunters covered all available escape routes while one adolescent male crept after the prey and caught it, whereupon the other males rushed up and seized parts of the carcass."[119] This is death by dismemberment, written in very polite terms.

So, if you thought the "empty hands" hunting we talked about at the beginning of this chapter may have been far-fetched, let's take another look from the perspective of anthropology. Here's what Dr. Craig B. Stanford of the Department of Anthropology,

University of Southern California has to say: "In spite of lacking the large canine teeth and tree-climbing adaptations that chimpanzees possess, [our most ancient ancestors] probably ate a large number of small and medium sized animals, including monkeys."[120] Chimpanzees are known for catching adult colobus monkeys by grabbing them and flailing them on the limb of a tree or on the ground. They often work as a team and push the monkeys to a treetop from which they can't escape. Stanford says that this model may be similar to what our ancient ancestors did and suggests that "the role that hunting played in [their] social lives was probably as complex and politically charged as it is in chimpanzees."[121]

What's Meat Got to Do with It?

In the descriptions of the killers above, there are many references to situations where an animal is killed and not eaten. There are many theories about why this may be so, and they may all be partly (or situationally) right. The kill may be opportunistic, and the predator will later return to feed. The kill may be a teaching opportunity, and although the prey will be eaten eventually, feeding is not the priority right at the moment. The kill may be territorial, not for feeding. The kill may be social (for example, killing the cubs), and as long as food is not in short supply, there's no particular reason to feed on this prey. Or the kill may be the result of an instinct that says, *When the prey runs, chase it* and *When the prey is caught, immobilize it.*

In Brian Palmer's article "Do Wolves Kill for Sport?," he points out that it's not just about wolves. Foxes often kill large numbers of chickens and eat only the head of each one. Weasels like to eat the back of the head and the neck of birds. Raccoons eat birds, too, but only the head and the crop.[122] We have seen feral cats eat chipmunks and leave only the heads; we've also seen them eat flying squirrels and leave only their little fur coats.

152 → On Hunting

Dolphins have been observed engaging in the seemingly gratuitous killing of porpoises—going so far as to use sonar to locate the victim's vital organs and increase the lethality of the strike—but experts haven't quite worked out their motivation. Some speculate that the dolphins use the porpoises for target practice, preparing for possible clashes with fellow dolphins who infringe on their territory.[123]

But even if we can't fathom the reason for a predator's kill, we at least understand that killing is what they do. A more mystifying question is, why do herbivores (vegetarians) kill? From the smallest to the tallest, there's a whole lot of killing going on out there. It's hunting of a sort...but it's not always for food.

- Prairie dogs are herbivores, yet they commonly kill ground squirrels, and often they are "serial killers." The females who were killers raised bigger litters and were in better shape than their nonkilling neighbors, "underlining the importance of eliminating competition for food while raising young."[124]
- The deer family are herbivores, yet they will kill when feeding, breeding, or security is involved. Deer have been seen eating birds. The bucks don't usually "fight to the death" on purpose, but they can certainly wound each other severely, and in the wild, this can be a death sentence. The bucks will also lock horns, and if they can't get separated, they'll die together. The bucks and does will fight over food, and the does will fight with each other (standing upright on rear legs and striking with their body weight powering the front hooves). The does will also strike at a predator to protect their fawns and will strike at an older fawn that is attempting to

suckle after it's been weaned (or one that attempts to mount the doe).

- Moose are larger and therefore more dangerous than white-tailed deer. They are also, when provoked, more aggressive. A cow or bull moose will charge a predator, a person, a snowmobile, a full-size vehicle, or even an oncoming train. If they make contact with a creature, they are likely to trample the victim and can kill dogs, coyotes, wolves, bears, and humans.[125]

- Elephants are beloved in the Western world. We have the idea that they are big "friendly giants" who are highly intelligent and very calm. The people who live in areas among elephants have a different sense of them. "We are used to seeing elephants as peaceful, even friendly animals, but they are actually among the most dangerous of wild creatures…Elephants can attack for several reasons; protecting their babies, their space, or simply, when they are in a bad mood. Males also enter a periodic condition known as musth, during which their hormone levels increase so much, that they go on a rampage attacking any creature they see, including lions, rhinos, and humans. For this reason, zoos and circuses often avoid keeping male elephants."[126]

- "The hippopotamus is the most dangerous of all herbivores in Africa; it kills more people every year than lions, leopards and crocodiles. Extremely territorial, the hippo, particularly the male, can weigh three tons or more, and has been known to attack both on water (even capsizing boats and kayaks) and on land, where it can run incredibly fast despite its fat appearance. It has the largest and mightiest jaws and longest canines of any mammal, and can bite an adult crocodile in half."[127]

- Rhinos have a well-earned reputation for being aggressive and ill-tempered. Perhaps because of their poor eyesight, they seem to have developed an attitude of "charge first, ask questions later." When it perceives a threat, the rhino will lower its head and charge in the direction of its foe. A two-ton rhinoceros galloping at up to thirty miles per hour is enough to scare most predators away. But if the opponent is still there, the rhino strikes with horn and teeth. Rhinos use their horns not only to defend themselves from lions, tigers, and hyenas but also in battles for territory or females. Males and females frequently fight during courtship, sometimes leading to serious wounds inflicted by their horns.

- The Cape buffalo is one of Africa's dangerous game animals, so called because of the difficulty and danger of hunting them. It is also known as "Black Death." The Cape buffalo is said to kill more people in Africa than any other mammal after hippos. It is estimated to kill more than two hundred people every year. Wounded ones are notorious for ambushing and attacking pursuers.

So, we don't really understand why herbivores (vegetarians) kill, but we know that they do. Although it's not for food, it is hunting of a sort. At times, it's more like our varmint hunting, and at other times, it's more like street fighting or gang wars.

What's Play Got to Do with It?

Child's play in mammals is where the youngsters learn how to survive: parental bonding, peer competition, feeding etiquette, and hunting skills.

Modern humans go through this cycle, but nowadays the hunting skills are often overlooked or transmuted to sports.

Instead of learning bush skills (like navigation) and animal lore (like tracking), today's youngsters are taught modern survival skills (reading, arithmetic, computer skills, culture) in an extended childhood that goes beyond the age of reproductive maturity.

Paul Shepard quotes Dutch anthropologist Frederik Buytendijk, who describes primate-herbivore play, like that of young gorillas who like to swing, hang, climb, tumble, slide, and wrestle, and carnivore play, more like that of lions or wolves, who stalk, rush, hide, creep, chase, elude, attack, and yield.[128]

The games human children play have deep roots. Our ancient background speaks vividly through the behavior of our offspring. The child who, in unfamiliar surroundings, stays near his parent's or grandparent's knee is demonstrating parental bonding. The youngster who steals his brother's cookie doesn't know he's part of an ancient lineage of humans who competed with their siblings for more food and attention. The mother who tells her daughter to wait for all the family to be seated at the table before she touches her food is teaching her child proper pack etiquette. And the parent who teaches a child to shop and prepare food is substituting these activities for the hunt.

But the need to hunt comes out in our children. As Paul Shepard points out, among the common games (especially of boys) is the dramatization of stalking and being stalked, killing and being killed. "At about the age of ten the boy's taste for killing and dying is uninhibited and unlimited…Adults, especially parents who are ideologically sensitive, generally misunderstand the importance of simulated violence and death in the lives of children who live in cultures where death is not a normal and accepted part of life and is hidden from them."[129]

Some children, of course, are still lucky enough to be taught to hunt, and the skills they learn will be covered in the next chapters.

Chapter Nine—Summary

Humankind has some remarkable similarities to the animal world when it comes to hunting. Hunting techniques used by predators in the wild also appear in various forms of hunting by humans.

One major difference in human hunting is that we strive to be "humane" in our killing. This is what truly differentiates us from predators in nature. Indeed, *humane* killing is what defines the modern *human* hunter. Thus, an ethical shot that will give a quick kill is always the goal. But Mother Nature can be a cruel mistress. In nature, predators have no reservations about munching away at their prey while it is still living and suffering.

In the wild, predators will happily gobble up any helpless young prey that they find, in many cases eating them alive. Modern human hunters, in contrast, try to preserve the young of game species. We often do not harvest females of the species. In fact, most hunting jurisdictions have controls around harvesting females and their young and allow it only where there is too large a population of these animals. It is the wise thing to do in preserving game for the future.

We strive to harvest our prey when they are old, past breeding, and likely to die during the coming winter (or dry) season. This is when they have the largest size, the biggest racks and horns, or the most distinctive mane. It is the right thing to do, giving a quick and ethical death at the end of the life cycle. We want to prevent the slow, painful death by old age in nature, which usually happens over days (if there is no predator to provide a quick death), as the creature is slowly eaten alive by insects and rodents. There is no end-of-life care in the wilderness.

Perhaps one of the best things we could learn from animal predators is to train our young to hunt. Some humans are blessed to have this kind of training as they grow up. They learn to venerate the wilderness through the hunting culture. And of course, we humans can teach our young to hunt humanely. Nature can be cruel—predators seem to torment helpless prey when they are teaching their young to hunt and kill. Human hunters seek to instill a code of ethical and humane hunting into the young hunter.

Chapter Ten

An Ancient Heartbeat

> Successful hunting, it could be said, is an act of terminal empathy: the kill depends on how successfully a hunter inserts himself into the umwelt of his prey—even to the point of disguising himself as that animal and mimicking its behavior.
>
> —John Vaillant, *The Tiger: A True Story of Vengeance and Survival*

It's easy to let our modern viewpoint (and sometimes arrogance) define our ancient ancestors, but on deeper reflection, it's clear that we stand on their shoulders. Their skills brought us to our current state, and although most of us no longer possess many of their ancient skills, we are deeply indebted to them.

We still possess the genes and hormones of these ancestors, and a few modern activities continue to demonstrate our lineage: breeding, feeding, and security (fight or flight). The same parts of the brain are activated, the same hormones are generated, and the same results are achieved. What isn't always so obvious to modern humans is that these three imperatives are also linked in our brains, and they each produce some of the same hormones. It makes sense. Breeding, feeding, and security all have to do with survival.

When it comes to sex, most people understand it as an innate drive. We wouldn't survive as a species without it. It shapes our lives. Many of our cultural habits and institutions form it and contain it. Similarly, most people understand security as a basic human need, and we design our institutions (especially government) to ensure it. Feeding is even more primal. It is required since, without sustenance, people would soon not be able to muster either security or sex.

Feeding includes hunting and gathering (and, more recently, growing and raising). We modern people usually think of hunting and gathering as separate activities, but it may well be that in ancient times, people would have thought of "feeding" as a single activity that included all means and all types of sustenance. (It's only a modern mind that struggles with the idea of whether fishing is hunting, for example.)

When we talk of hunting in this book, we are specifically thinking of activities that result in feeding on other animals. The skills required to do this have not changed much over time, although the tools have. The similarity of the skills modern hunters share with our ancient ancestors bonds us to them. Today's hunter feels a special relationship with hunters through time, and this causes us to believe that the bond is primal and significant. Indeed, it is the yearning to connect deeply to our past that draws many modern humans to the hunt. The hunt is in the tissue of our being. The skills of the hunt are the skills of our ancestors, and by acquiring and using these skills, we seek to understand our past… and our present.

Inside of each of us is the ancient brain and the endocrine system of the hunter. To feel whole, we need to see him, acknowledge him, and celebrate him.

There are seven separate components of the hunt that have not changed [preparation, search, stalk, kill, butchering, cooking, eating]. This leads to some similarity in the philosophy of both the ancient and modern hunters.

—Carlos Gallinger, *Hunting and the Human Way*

Campfire Story: Viv's First Deer

It was the last Saturday before bow season for deer was officially over. After about forty-five minutes, Keith pointed out a buck about one hundred meters to our left. As it walked along the tree line on the other side of the clearing, it appeared to be limping. As soon as we realized it was lame, I knew this was the buck I wanted to get. Unfortunately, while still out of bow range, it disappeared into the tree line. Since it was approaching sunset, we began to think that we weren't going to see any more deer.

As 5:00 p.m. came, we watched a doe come out of the tree line to the left of our stand and make its way to the carrot pile.[130] As this was happening, two more emerged from where we had seen the buck earlier, and another one from across the field. Seconds later, the same buck came out from where it had disappeared. By this time, I think Keith and I both lost track of how many does were there. We were both thinking, "Oh boy!" As I got into position with the bow, we both kept our eyes on the eight-point buck, limping its way to the carrot pile. As soon as it got within the thirty-meter range, Keith told me to go ahead and take the shot whenever I was ready. I kept my eye on the spot just behind the right shoulder and pulled the trigger.

It's hard to remember exactly what happened after that, and I'm sure Keith would say the same, because we were both so excited. (Be prepared for a rush of adrenaline.) The deer died a few minutes after it was shot, and we found it not far from where I had shot him. Linda came out to help us, and as I remember her saying before, it's amazing how it takes no time at all for a living animal to become meat. I watched Keith gut the deer and later helped him skin it, but I don't think I processed what happened until I got home later.

Looking back, I feel as though that was the deer I was supposed to shoot. It had a broken leg from some sort of accident and had no fat on him at all. He wasn't interested in the does at the bait pile; he was only looking for food. Yes, it is a sad thing when an animal dies, but it is even sadder if it suffers through this process. This deer that I waited weeks for has granted my family and me a freezer full of organic meat, and I am so grateful for this and every single thing I learned and experienced through the entire hunt. Overall, hunting is an amazing time, and I would rate it 12.5 out of 10.

The Hunter's Eyes

> Focused alertness towards one's environment is fundamental to the hunting experience and a central aspect of its appeal. Yet not only does hunting stimulate a hyperawareness in the hunter of her surroundings, it also inspires an intense sense of intimacy between the hunter and the prey.
>
> —T. R. Kover, "Flesh, Death, and Tofu" in *Hunting— Philosophy for Everyone*

One of the most interesting aspects of hunting is developing "hunter's eyes." This is more than just a sharpness of vision; it is a mental skill that enables the hunter to see things in a way that can't be described simply as vision. It is in part sensitivity…seeing the details of what is there, the details that untrained eyes tend to sweep over. It is in part patterning…recognizing what is "normal" and what is different from the usual. It is in part attribution…

attaching special meaning or depth to what could otherwise just be a nice view.

Coauthor Linda's uncle Austin used to take his daughter out hunting with him in the bush near their farm. They would wander along the trails and enjoy the peace and depth of the woods. After they returned home, he would ask her what she had seen. She started describing the wildlife and the flora and the rockeries. "Yes, yes," he confirmed, "I know what you looked at, but what did you *see*?"

> Labeling what was around me connected me to where I was.
>
> —Hugh Brody, *The Other Side of Eden*

Modern, so-called primitive, hunter-gatherer tribes that still inhabit our earth have been thoroughly analyzed in many ways by modern socio-anthropologists and biologists and other scientists. The findings vary, but generally speaking, we could learn a thing or two from these "primitive" people!

One of the things most hunter-gatherers can do that we modern folks cannot do is name the elements in our environment... plants and animals...and specify many details about their characteristics and importance. In fact, the way that hunter-gatherers organize their knowledge about plants and animals is complex...as complex as any other classification system that a modern, urban, educated human might make to describe his world.

Movement

One of the aspects of the hunter's eyes that is vitally important to both predator and prey is the sense of movement. An animal may blend into the background beautifully, but the moment it moves, we see it. This is part of human vision that is not well studied but one that every hunter knows—from both sides! A deer can blend into the surroundings, and we can't see it, and it's capable of moving so slowly that it's almost beyond our ability to perceive. But the moment it flicks its tail, we jolt to attention and focus on that spot. And from the other side…we all know that if a deer sees us move, we're busted!

It's likely that this is due to the visual processing part of the brain, which is very sensitive in the deer and will be very sensitive in you when you've developed your "hunter's eyes." One research project has studied a specific brain area that reacts selectively to movements in our surroundings and that reportedly "reacts to movements around us regardless of whether or not we follow the moving object with our eyes."[131] This is likely an ancient skill, deeply embedded in our hunter's brain.

Night Vision

One aspect of human biology that's quite different from many of the animals we hunt is night vision. Ours is extremely limited. We lose color vision and depth perception, and some researchers believe we also lose some of our movement processing (i.e., we perceive objects as moving slower with our night vision than they seem to move in the daylight). Of course, some animals (both predators and prey) are fully nocturnal and have particularly good night vision though it's thought that they sacrifice color vision and sharp focus. We can well believe that our ancestors, having moved from the nighttime security of the trees, would seek out caves and eventually use fire and dogs to help protect themselves from nocturnal predators.

Animal predators (including humans) have their eyes set at the front of the head, enabling binocular vision (and, therefore, the depth perception important for the hunt), while prey animals have their eyes set at the sides of the head, enabling panoramic vision (and, therefore, threat perception important for the escape). Many hunters believe that if you look at the game with both eyes, it senses you're a predator, but if you look at the game sideways or use small, squinty glances, it won't notice you as easily or won't fear you when it does.

Patterning

Another mental aspect of vision is patterning. Our ancestors surely were masters of this skill. The modern hunter uses patterns to figure out where to find game (for example, finding trails, waterways, bedding areas, feeding areas) and then to see game even when it's camouflaged. (For example, seeing a horizontal line in a wooded area usually means a deer or moose is standing there.) When we went on safari in Namibia, it took us several days to develop our animal and environment "templates," imprinting the images into our brains. The hardest animal to see was the kudu because of its coloration, its practice of putting cover between itself and the hunter, and its habit of moving very slowly toward and away from (rather than across) the hunter's field of view.

Attribution

An important aspect of the "hunter's eyes" is attributing meaning to what we see. When a hiker sees an oak tree, for example, he may identify it as a landmark to return to. Or when a naturalist sees an oak tree, he may look at its bark or leaves to identify the specific type of oak and the overall age and health of the tree. When a deer hunter sees an oak tree, he immediately looks for acorns on the ground and then game trails leading into the area. These attributions are what make the tree meaningful to each individual.

Body Language

A key part of the "hunter's eyes" is interpreting the body language of the game animal. This is a skill many people lack, even when it comes to understanding the body language of the people and pets they see every day. One of the standard body language postures that most dog owners know is the "play" posture, where the dog lowers his chest near the ground with his front paws somewhat extended. It's usually accompanied by tail wagging, ears up, and mouth open. Dogs have several other common postures, indicating alertness, offensive threat, submission, and so on. The best body-language observers start to see the details of this communication and can often predict the pet's next move based on a small indicator.

The hunter watches body language whenever he has an animal in his view, even if it isn't the animal he's hunting. If he's been in the North American bush enough times, he has watched territorial red squirrels, pouncing foxes, and stealthy hawks. If he hunts deer, he has watched many that he did not take. For example, he's seen the doe and fawn nuzzle a greeting, the dominant doe challenge the lesser doe, bucks in rut put their heads down and engage in some sparring. He can tell by body language whether the animal is nervous or relaxed, whether it's going to stay and feed in front of him or if it's going to bolt at the slightest excuse. The hunter can tell a buck is on the trail of a hot doe by his purposeful step and nose-to-the-ground focus. And when he shoots an animal, the hunter watches body language that confirms an effective hit. He knows whether he needs to follow up the shot or if he can just unload, pack up his kit, and take his time to start tracking.

Developing "hunter's eyes" takes time in the field, and not just hours, but attentive, engaged hours. There are modern tools that will help your vision (binoculars, optical sights, night vision), but it is only through mindful practice that the hunter develops the mental skill.

Marksmanship

Staying proficient at shooting is a lifetime activity. The ethical hunter will constantly hone his or her shooting skill and retain his or her shooting confidence. An ethical hunter has the responsibility to know the capability of the weapon used.

—Jim Posewitz, *Beyond Fair Chase*

Hunting marksmanship is the art of placing lethal force in the most effective location to ethically dispatch a game animal. This implies that the hunter understands the lethal force he's using (whether it's a spear, an arrowhead, or a bullet) and that he has what it takes to use it well. In most definitions of marksmanship, only guns are included; however, there is certainly marksmanship in any projectile that's launched.

If you are going to become a dedicated hunter, we believe you should have three rifles. These rifles should be of the same make and model so that they are identical in every way except caliber. This way, the handling drills you learn with one are transferable to the others. You want to be able to move from one rifle to the other and know the safety is in the exact same place with the exact same feel. We chose the bolt-action rifle because of its simplicity, reliability, and inherent accuracy. The shot should be the easiest part of the hunt.

The calibers we chose for the first two rifles were .223/5.56×45mm and .308/7.62×51mm. We chose these for a couple of reasons. First, the .223 is an excellent cartridge for small to medium game, thus providing us with valuable hunting practice and training. The .308 is an excellent caliber for medium to bigger game. Both calibers are used by many militaries throughout the world. Therefore, you can find lots of military surplus ammo available at a very reasonable cost. You can use this ammo on the range for practice and training so that you can get good with your rifle. You would *not* use this ammo for hunting. When the season is approaching, you would zero your rifle with the ammo you plan to use for hunting.

Our third rifle is one we use for longer range shots or for bigger game.[132] Now is a good time to get one of the magnums. We went with the 7mm Winchester Short Magnum because it produces excellent long-range energy, has minimal recoil, and is easy to shoot well. This rifle has worked well for us while we hunted a variety of animals, from springbok to eland, zebra, red stag, and whitetail at ranges out to 700 meters. On our most recent trip to Africa, Linda used her .308 on springbok at ranges in excess of 300 meters, and our friend Carl shot zebra and oryx at ranges just under 300 meters with his .308. Shot placement trumps power every time.

You must do everything possible to ensure shot placement is right every time. You must practice and train so that the correct firing of your rifle is something you can do subconsciously—that is, without thinking about it. The rifle should fit you properly so that you can mount it smoothly without it catching under your arm or on your clothing. (Think carefully about this one—in many parts of North America, you train in the summer wearing a t-shirt, and you hunt in the fall wearing a jacket.) You might have to get the butt shortened so this all works for you. Practice so the safety comes off at the right time when bringing the rifle onto target without having to think about it. Be smooth with all your handling drills.

Keith vividly experienced the benefits of practicing being smooth during a recent deer hunt. He was using a wonderfully comfortable hunting hut, overlooking a pile of carrots used as bait, at a range of fifty meters. He slid open the windows and decided that he would use the windowsill as a rest to steady the rifle and practiced being smooth in this shooting position. He was sitting in a nice chair, reading a book, and enjoying the solitude of the bush. He relaxed into an enjoyable wait, looking up after each paragraph to see if anything had arrived.

After several hours, he looked up and saw that a big-bodied buck had just arrived, clearly looking for the does that frequented the carrot pile. The antlers had only six points but were massive. He immediately made the decision to take this buck. He placed his book down on the floor of the hut, and it made a slight noise. The buck immediately turned its radar ears in Keith's direction, and its body language said that it was clearly ready to bound off. Keith could definitely see he had to act fast but quietly. He doesn't remember reaching for the rifle, mounting the rifle, or removing the safety, but he vividly remembers seeing the sight picture in detail. The rifle fired, and he could see the hair ruffle where the bullet struck—clearly a fatal shot. It was all over in seconds.

One of the important points about firing a good shot is recoil management. The recoil must be the same for every shot, or the barrel will be pointed in a different place as the bullet leaves the gun and will, therefore, land at a different place from where you are aiming. If you're in a jurisdiction that allows suppressors for hunting, they can take care of some of the recoil and some of the noise.

For the average hunting rifle, the best sight would be a low-powered optic with a simple reticle. When you're expected to use equipment under stress and on demand, the simpler it is, the greater the chance you will use it correctly. Optic sights can provide better results (than iron sights) under low light situations, such as early dawn and late dusk, or in wooded areas with a thick canopy.

Ethics and Energy

The best place to hit an animal that will guarantee it will die is one that puts the bullet through both lungs. This is called "double lunging," and there is nothing on the face of the earth that can survive such a hit. Although it doesn't always cause the animal to drop in its tracks, it destroys the least amount of meat, uses the largest lethal zone on the animal, and is always a fatal shot.

There are good reasons to hit an animal in the front shoulder or high in the shoulder. Our professional hunter (PH, a hunting guide) in Africa wanted us to hit our animals in the shoulder because saving meat was of less importance than dropping the animal where it stood. This can especially apply when hunting dangerous game. A lung shot, even a double-lung shot, can give the animal ten to twenty seconds to charge or make it into thick underbrush, where retrieving it can be a little too "exciting." The idea here is to immobilize the animal by breaking one or both shoulders so that its front legs stop working.

There has always been a debate in hunting circles about what caliber and what bullet type is best. But as Jim Carmichel wrote in an *Outdoor Life* article, there could be another "knockdown" factor that we can't see. Apparently, some veterinarians were engaged to cull buffalo. After shooting each animal, they recorded their immediate behavior, and then they dissected each one to see what damage the bullet had done.

When the brains of all the buffalo were removed, the researchers discovered that those that had been knocked down instantly had suffered massive rupturing of blood vessels in the brain. The brains of animals that hadn't fallen instantly showed no such damage.

Their conclusion was that the bullets that killed instantly had struck just at the moment of the animal's heartbeat! The arteries to the brain, already carrying a full surge of blood pressure, received a mega-dose of additional pressure from the bullet's impact, thus creating a blood pressure overload and rupturing the vessels.[133]

We can't see whether our target animal is between beats when the bullet hits him, but at least we might have a better understanding that there's more to knockdown power than meets the eye.

When it comes to bullets, it used to be that bigger was better. That was so that as the bullet penetrated and broke up, there was still enough mass to continue penetrating and creating tissue trauma. You might have started off with a 200-grain bullet and ended up with 100 grains. However, the bullet makers of today know how to make a bullet expand without breaking up while retaining most or all its weight. You can start with a 150-grain bullet, have it expand and create massive tissue trauma, and still end up with a bullet that weighs 148 to 150 grains.

The advantage of using this lighter bullet is that there is less recoil, flatter trajectory, less time in flight, and more energy. Kinetic energy is a result of mass and velocity, with velocity being the more significant factor. (In the formula used to calculate kinetic energy, we use half the mass but the square of the velocity.) It's hard to say just what amount of energy is required to efficiently take down game. Some states in the US require a minimum of one thousand foot-pounds at one hundred yards before the cartridge is recognized as legal for hunting. A similar standard is applied in many areas in Africa. When we hunt at long distances, we like to have something close to one thousand foot-pounds of energy at the distance we intend to shoot.

The way the bullet is designed to perform on contact with the animal is also important. When hunting coyotes or other relatively small, thin-skinned game, you must use a bullet that is designed to expand very rapidly so that it's creating the most tissue trauma at exactly the right place as it passes through the lethal zone. If you are hunting big, thick-skinned, big-boned, heavy game, you will want a bullet designed to penetrate well into the vitals before it expands. Modern hunting bullets are the best they have ever been, designed specifically to do what is required of them. Pay attention to what the ammunition manufacturers say about their product. In this regard, they know best.

The standard to achieve in your training and practice is to be able to deliver an ethical shot, every time, onto the animal you plan to hunt. For deer, this could equate to being able to hit a paper plate. You must know your limitations—at what distances and from which positions can you consistently hit this paper plate. Then, once in the field, set yourself up for the distance and position from which you know you can deliver that ethical shot.

Learn How to Shoot

Marksmanship is not something magical, nor is it a skill you are born with. It's something you learn and then practice just as you might practice your golf swing. The fundamental skill is hand-eye coordination, just like your golf swing or throwing a ball.

During Keith and Linda's marksmanship courses, we teach five marksmanship principles:

1. Position and Holding Pattern: The shooting position must be firm enough to support the required holding pattern. The holding pattern is the amount of movement you see when looking through your sight at a target. For the deer hunter, it must be inside the "paper plate" on a deer, the size of the vital area. To make your area of aim smaller, modify your position or use a rest.

2. Natural Body Alignment: The rifle must point naturally at the target without lateral forces or muscle tension. If you apply extra force, the rifle will move in that direction during recoil. You must be relaxed when firing a shot, using only neutral force and only enough to control the rifle.

3. Sight Picture and Breath Control: You must establish the correct sight picture and control your breathing to minimize movement. Choose your point of aim to get the correct point of impact for the situation (point-blank, holdover, moving deer, or windy conditions). Make sure you have a full field of view in the scope. When it's time to take the shot, hold your breath. If you're very excited, take a little extra gulp of air to settle yourself down. If you're really excited, do combat breathing.[134]

4. Mental Program: The mental program is a series of conscious thoughts leading to the subconscious firing of a shot. It's a checklist. Once you've learned how to fire a shot correctly from any position, the firing of a good shot is 90 percent mental.[135] This mental program, this checklist, can include any of the steps you think are important. We use the following: "Breathe and relax"; "Correct target" (point of aim); "Level" (important for long-range shots); "Sight picture" (or sometimes, "Float, squeeze, and wait" or just "smooth").

5. Trigger Release and Follow Through: The whole objective of the trigger squeeze is to do it without disturbing the sight picture. Put enough finger on the trigger to apply a slow, smooth, steady, subconscious squeeze, straight to the rear. Since your primary job during the firing of a shot is to watch the sight picture, then the follow-through should be easy. You must be able to "call the shot" by being aware of where the crosshairs were when the shot was fired.

Practice

One of the best ways to practice your marksmanship is by dry firing. This means going through the complete shot procedure without the use of ammunition. The shot procedure, whether dry or live, is the same. You are trying to get the correct procedure imprinted on your subconscious mind so that you will do it this way under stress and on demand.

Reading the Wind

You need wind-reading skills to be an effective shot at long ranges. There is no way that an instructor can knight you with the magic cleaning rod and instantly make you a master at reading the wind. You can certainly learn some basic techniques from an instructor or a good book on the subject,[136] but then you must go out and practice shooting long range in the wind.

> To hit the target we need good judgment, experience, deliberate thinking, and skills developed by habit. Indeed, at the heart of hunting is a discipline that requires developing those skills unique to hunting. You must cultivate the skill of the kill: unflinching, steely nerves; tracking and orienteering abilities; and, of course, superior marksmanship. This is built on hard work.
>
> —Jesus Ilundain-Agurruza, "Taking a Shot: Hunting in the Crosshairs" in *Hunting—Philosophy for Everyone*

"That Guy"

You don't want to be "that guy." During Keith and Linda's Hunter Marksmanship and Safari Marksmanship courses, we hear it all, every excuse for missing or wounding an animal. We also hear all about how the animal was brought down with one shot but only because the animal was big and unlucky. We can't bear to hear a hunter say, "I can't hit paper, but I always get my deer." The fact is, if you are a poor shot on paper, you are a poor shot in the field.

We train many marksmen, and the ones we're truly proud of are the ones who understand their capabilities and limitations and apply them to an ethical shot on game. They get the right equipment, get it fitted to them, learn how to use it, and practice with it.

Campfire Story: I Got an 8-Point Buck This Morning!

I thought I was going to be caught again this year, what with only seeing antlers and a bit of his head. He would put his head down and bring it back up and look right at me. I could have gone for the head shot, but a slight miss and it's the jaw shot. I have now tracked two deer with shot jaws done by other hunters in our camp, and I don't like that.

I had the scope on him, but I couldn't get his shoulder in the cross hairs. So, for twenty minutes or so, the game went on. My heart got to racing, and I said to myself, "You gotta calm down," and I did the combat breathing. It really helped, even with the muscles shaking from the position; it helped them too. I tried cranking the scope up to full power, hoping I'd find a hole in the bush, but still no shot. Finally, he moved, but it looked like he was going to move back into the bush. The shoulder appeared, but suddenly the cross hairs started to go jumpy, and this voice spoke in my head, "Breathe and fire the perfect shot." Now, where do you think that came from? It was just like in *Star Wars* when Luke is told, "Feel the Force." Keith, your middle name isn't Yoda, is it?

The crosshairs settled, the gun went off, and the deer was gone. The shot went off right where it was supposed to, but no deer. I was able to call the shot, and it was good. So, I leaped out of the stand and ran to where the deer had been standing. And there it was. It had fallen into a depression and was out of sight.

Looked like he never felt a thing. The bullet went in just behind the right shoulder, broke two ribs, and stopped on the skin after going through the left shoulder. I was using my .308, at seventy-seven yards.

It was amazing when I heard myself say, "Breathe." It was just like autopilot. Thanks for the coaching and training. It made for an amazing hunt.

Field Craft

Keith was first introduced to field craft as a kid growing up on a farm. He didn't know it was called something; he just thought it was "sneaking up" on something, usually a groundhog, to get within range with his old .22 caliber rifle. He noticed deer tracks in the plowed fields or along the edge of the creek. The size of the wolf tracks around the fresh-killed livestock would always send a chill down his spine. Older members of his family would show him the best place to be during the deer season—like a pinch point in the tree line between two hay fields. Or, they said, look for the well-used runways created by snowshoe hares and be nearby it when the beagle was turned loose. It wasn't until he joined the army that all of this was called something: field craft.

Good hunters, those who always seem to be in the right place at the right time and always get their deer, have developed the ability to "read sign." You need to do more than just see a track and recognize what made it. You have to tie it in with other tracks and other sign to get the whole story. Knowing the way the game lives, recognizing the terrain, and linking those to the other sign can tell you if the animal was passing through or lives in the area. Knowing how their habits change with the seasons and applying that to what you see on the ground will tell you if this is going to be a good place on opening morning.

> There is one illusive mode of hunting…the search for scats, tracks, feathers, dens, roosting, rubbings, dustings, diggings, feedings, fightings, or preyings collectively known to woodsmen as "reading sign." This skill is rare, and too often seems to be inverse to book learning.
>
> —Aldo Leopold, *A Sand County Almanac*

Getting Started

Whether you are hunting on familiar ground or going in for the first time, the best way to get an overview is with a topographical map and an aerial photograph.[137]

With some training and practice using these together, you can get a very clear picture of the land features you are looking for and then can go there to look for signs of activity. Your map and photo will show tree lines and clearings and how they might connect to make pinch points. They will show ridge lines and draws, streams and beaver ponds and lakes. Now you will know where the best land features are, and you can go there in person to see what the sign tell you.

It's fun to learn to read a topo map, and it's a real confidence builder to be able to navigate across country with map, compass, and GPS. Camping out overnight in the wilderness can always be an exciting adventure, and you can practice many survival skills.

When you find the places with a concentration of sign, put up a trail camera and get pictures of what is making those tracks and at what time of day they are there. The best trail camera is the kind that will send you the pictures so you don't have to disturb the area by going in to retrieve the chip. If you do have to retrieve a chip, make good use of this intrusion by looking carefully for recent activity. Look for tracks but also look for scat, rubs, and scrapes and for trails coming and going from bedding areas to feeding areas. Know what the game likes to eat and find concentrations of this food or provide something they like with baits (where legal). Your trail camera will tell when it's the best time to hunt an area based on the activity.

> Because I hunt, I know what my local whitetails are likely to be feeding on with each change in season; I know how to read tracks and picture each footstep of the animal at the end of them, to learn something of its size and conditions through telltale strands of hair or fur rubbed into rough bark, to decipher the daily diary of feces, and—yes— to translate the bright dribbles and spews of a blood trail.
>
> —Bruce Woods, "The Hunting Problem" in *A Hunter's Heart*

Tactics

Next, you should figure out how you're going to hunt this area. It is difficult to simply "sneak and peek" through the bush and get close to something. Certainly, it has happened. Sometimes, you can get close to a buck in rut when you catch him focused on a scrape or rub. But if you must move over ground covered with dry leaves, each step sounding like you are walking on corn flakes, the advantage goes to the game with their radar hearing. In this kind of terrain, you are better to be still and let the game make the noise coming to you. So, you need to figure out the best place to put a stand or blind.

The ability to prepare your hunting area with hard stands and bait piles may be limited if you're hunting public land, as many Americans do. But if you have your own land or have long-term hunting permission, preparing your hunting area can be a year-round pleasure.

Operations

No matter the style of hunting that you select, you must always keep wind direction foremost in mind. Your usual stand location may not be the best if the dominant wind is blowing past you to where you expect to see the game. There are lots of cover scents and scent-blocking clothing that you can buy, but in the end, it is hard to make a human not smell like a human to a game animal with a nose that has developed through thousands of years of evolution. The only sure way to beat such a nose is to have the wind blowing from them to you. You might consider having more than one stand location covering the same area so that you can always have a place to go, no matter the wind direction. Even a tree stand or a high stand won't necessarily block your scent because downdrafts can carry your scent down to the height of a deer's nose.

You should also think of the wind when selecting a route to your stand location. It is illogical to spread your scent throughout the woods that you are hoping will produce your game. You may have to plan a couple of routes to accommodate wind directions.

Keith was once hunting along the side of a maple ridge and had taken up a position against a large maple tree about halfway up from the floor of the valley. The wind was blowing quite hard crossways on the ridge. A small doe appeared from the valley and started to make its way up the side of the ridge, directly toward Keith's position. He wasn't hunting does, but he was concerned that the deer would eventually alert on him, sound an alarm, and make lots of noise getting out of there. But the deer stayed upwind to him and came closer. Eventually, it was standing on the other side of the tree from Keith, now within arm's reach. He couldn't resist and reached out and touched the deer, startling the poor animal beyond comprehension. It made all the appropriate alarm calls and lots of noise. (He says it was worth it.)

Applying Your Skills

The spot-and-stalk style of hunting is perhaps the most challenging and satisfying. This works best when you can spot the animal from a distance while it is unaware that you are there. You then must plan a hidden route to within range of the animal using camouflage, concealment, and different methods of movement, depending on the terrain. Above all else, you need to have the wind in your favor by picking a route that keeps the wind in your face.

> A whole different metabolism kicks in…I crouched periodically to look…[T]here were light tracks and sign everywhere. I moved one step at a time, placing each foot in between forest litter, eyes shifting from ground to foreground, straining…I moved individual branches to one side and weaved slowly forward so as to avoid the sound of needle or twig on fabric. After twenty minutes and two hundred feet…I froze. The odor was strong. The hair on my neck stood up.
>
> —George N. Wallace, "If Elk Would Scream" in *A Hunter's Heart*

Once the correct shot has been made, then the work begins. You now need to recover the animal, field dress it, and get it out of the bush where it can be properly butchered. If the animal drops where it was hit, then recovery is easy. But sometimes, even when hit properly, the animal will go a distance. In bush country, a deer can disappear in about twenty yards. The best way to find it is to follow the blood trail. If hit properly, the animal will be bleeding from its wound or from its nose or both. Take your time, be patient, find the blood spots, and go cautiously from one to the next until you come to your game.

Field dressing is a skill all its own and will need one-on-one teaching from someone with experience. Before you skin the animal, you need to think about the trophy and how you might want it mounted. Butchering is another skill that takes training and practice, and unless you have the knowledge, you may be better off taking your game to a commercial butcher.

> I feel the smooth, warm lobes of liver and remove them. I strain my thirty-six-inch reach all the way up to his diaphragm to cut the peritoneum that will release this mountain of viscera. The jays are already here. Half inside, I find the heart, cut it free, and cradle it in my hands. Grandpa Gallegos, more Indian than Spanish, used to cut off the tip of the heart and eat it at once. "This gives you the strength of the animal," he had said.
>
> —George Wallace, "If Elk Would Scream" in *A Hunter's Heart*

Cooking is the final skill that is required. The cooking can make or break the enjoyment of the hunting experience. There are many wild game cookbooks available (and more and more online recipes and videos), or you can find that person whose wild-game cooking you've enjoyed and get them to show you how it's done.

Campfire Story: Sylvia's First Deer—A Textbook Shot!
That afternoon I arrived in my blind at 2:45. I went through the usual steps of loading my rifle, placing it on my shooting stick to check to see if it was positioned to make a comfortable shot, and getting out my ear defenders. I opened my reading book to where I had left off, and the first time I looked up, a deer had snuck in and was feeding on the carrot pile. I noted quickly that it was a buck, so I laid down my book, put on my ear defenders, and, when the deer's head went down to pick up another carrot, slowly picked up my rifle and placed it on my shooting stick. I took off the safety; I was ready to fire. When his head came up, he stood facing me and stared in my direction. At this point, I wasn't sure whether I had been busted, but I knew I had to wait for the right shot. I waited; he watched. I was only thinking of making the right shot. I knew I had to wait for him to turn. I had him in my sights; I knew where

I wanted to place my shot. I sat there for ten or fifteen minutes before he turned. I placed the crosshairs, and the gun went off. I guess I pulled the trigger. Honestly, I did not feel any recoil.

It was 3:30 p.m. The deer jumped and ran off. I was excited, but surprisingly, I did not find that the adrenaline rush left me out of breath like you see on all the TV shows when the hunter makes his kill. I texted my husband, who was sitting on his watch at a nearby hunting camp, saying, "I shot a buck, and I know he is leaking!" When I looked where I took the shot, I noticed that as the deer ran off, his track looked like he was stumbling. Wanting to let the deer lie down, I went back to my blind and located my tag. I was sure I had killed this buck.

Word went out quickly. By 4 p.m., I had help to track the deer. It was exciting when we first spotted blood because it confirmed my belief that I had made the kill. At dusk, we found him about twenty yards from a pond. This was within one hundred yards from where I took the shot.

It took one shot with a 150-grain bullet. At the clinic, I learned it was all about making a shot I was confident I could make with a gun that was properly zeroed before the hunt. Now I know how good it felt to take my first animal. This deer is my first kill. My first comment was that it was a textbook shot. Thanks, Keith, for helping me learn how to get myself and my gun ready for the hunt (I did have my gun fitted and put a muzzle brake on my gun after the clinic).

Linda, I'm hoping you'll share your venison curry recipe. Nothing tastes better than your own deer.

Other Friends of the Hunt
Another style of hunting is with the use of dogs. Hunting with dogs has a historical significance as it is likely one of the main reasons why early man domesticated the wolf. There are certainly more skills required to raise and train your hunting dogs.

A tremendous bond develops between you and your dogs as you need to work with them almost daily to ensure they are properly trained. Hearing your "strike" dog make first contact with a scent trail and hearing the rest of the pack join in is exciting. Once you're experienced with your dogs, you'll hear the entire hunt develop through the different "baying" sounds. You'll learn to recognize each dog by its sound and enjoy the excitement of trying to keep up with them. But the training of these hunting dogs is not for the novice to do on his own since a good dog can easily be ruined with bad handling. You must have experienced help to get the most from these hunting companions.

In some hunting zones, the use of a horse is the logical choice. As with the dogs, these horses need to be trained and handled by those who know how. There is historical significance here as well since early man domesticated the horse to carry loads and to transport men. In addition, horses were used, along with dogs, to run game down to where the rider could kill it with a lance or spear. Since man could not always outrun his game, the horse accomplished that for him. One can only imagine the excitement of the chase on horseback. Alison Acton wrote that riding to hounds involves a "visceral engagement with a living landscape," and, she observed, the best horses are the ones who really love it.[138]

Where allowed, the use of motorized vehicles is another consideration to aid in your hunt. All-terrain vehicles (ATVs), four-wheel drive pickups, small airplanes, or helicopters are sometimes required to get your game out of the bush and to the butcher. In almost all hunting areas in the world, it is illegal and considered unethical to take down game while using a vehicle.[139] These vehicles can be used to get you to your hunting area and can be used to cover large areas in search of game, but once you spot your game, most hunters use the stalk (on foot) to get close enough for the ethical shot.

Physical Fitness

The more fit you are, the easier the physical aspects of the hunt will be. Certainly, physical capabilities differ from one person to the next. Mobility-impaired people can still enjoy an ethical hunt. Keith and Linda have a friend who is quadriplegic and who hunts deer every year; he has a stand where he can sit comfortably and rest his rifle, ensuring a good clean shot.

You need the strength to carry a pack and a rifle in the field as well as to bring the meat and trophy out. Aerobic fitness is especially important if you are hunting in high altitudes or face rough terrain like rocky areas or sand dunes. You also need aerobic endurance to be able to control your breathing when it's time to take the shot. Flexibility helps reduce the chance of injury and makes it easier for you to stay still for long periods. It's also important for being able to assume a comfortable, well-balanced, and relaxed shooting position. Depending on what area you hunt, you may need to deal with extreme cold or extreme heat. Stamina is important in hunting and can be enhanced by your physical fitness.

Hydration and nutrition are important too. While most of us can easily go a day without perfect hydration and nutrition, it's hard to stay focused on the game if we lack either. And lack of fluids can easily impair both vision and judgment, two skills we must have to complete the hunt successfully. If you travel internationally to your hunt, you'll also need to make sure you've taken care of any inoculations you might need to keep yourself healthy in that environment.

Unquestionably, some hunts may require more fitness than another. One of the hunts that requires more fitness than most is the high-mountain sheep hunt, where the hunter must deal with both difficult terrain and thin air. You should evaluate your capabilities honestly and select a hunt that's within your ability. And, if you're lacking in some way, you can usually do something about it! In fact, many hunters find the need for hunting fitness is a great motivator to get them to the gym.

Mental Toughness

Alertness is an especially important part of the hunt. Having said that, we must acknowledge that one of the most enjoyable parts of any hunt is the nap. The hunter's nap may last only minutes, but many hunters agree that it is the most restful shuteye we have ever had. There's just nothing like the peace of being in the bush and totally surrendering to our natural environment. Most of the time, though, the hunter is alert and engaged. In hunting from a stand, there may also be long hours of waiting, during which time most hunters read, do puzzles, or play on their electronic devices. Most hunters have a routine that ensures they regularly (and frequently) look up and intensely examine every detail of their area.

While mental endurance is required to stay alert, being engaged and immersed in the surroundings is extremely restful for the mind. The cares of the world fade, and the business of modern life becomes irrelevant, if only for a few hours. Concentrating on a natural environment taps into a peaceful inner core of our being.

> In the woods or on a stream, my concentration is so intense that for long periods the rest of the world is almost forgotten. I also immerse myself in books and magazines to acquire an understanding of nature, plants, insects, birds, fishes, and mammals, and the complicated interrelationships among them. I will read almost anything on these subjects, they interest me so much.
>
> —Former US President Jimmy Carter, "A Childhood Outdoors" in *A Hunter's Heart*

Another aspect of the mental toughness required for the hunt is dealing effectively with fear and excitement. In our

Campfire Stories throughout this book, dealing with the emotions of the hunt was a key part of each hunter's success and satisfaction. The hunters all relied on their preparation to ensure they would be able to handle the emotional roller-coaster. They stayed intensely focused on the job they needed to do. And one hunter mentioned using combat breathing to calm himself during the final moments.

If you do only three things to improve your mental toughness, make it these:

- Prepare yourself in every way, in every detail.
- When the moment of truth arrives, focus on your job.
- When it gets intense, do combat breathing.

One of the reasons new hunter Viv (who told one of the earlier Campfire Stories) was so well prepared was because of the Hunter Safety course she took. (This course is required to get a hunting license in her jurisdiction.) As part of the course, she watched a video of the field dressing and skinning of a deer. She knew what to expect, so the real thing didn't trouble her at all. She told us later that her only concern was what it might smell like. She was surprised that there was really no smell at all. Her arrow had produced a good clean kill, so the guts were intact. And the buck was not really rutting, so there wasn't even any musk smell.

Coauthor Dave tells his military and law enforcement audiences that the best way to prepare for combat is to hunt. One of the things the hunter learns is an intimate familiarity with the sight and smell (and even the feel) of what is inside a living creature. If the first time you encounter that stimulus is some horrible crime scene, terrible combat zone, or tragic accident, then you have been set up for failure. The hunter knows these things and knows them with a powerful positive association: they mean a successful hunt and food on the table. This is called *stress inoculation* and is a critical factor in reducing the possibility of a post-traumatic stress response. Although this is essential for military, police, and, really, any first responder, it can be useful for anyone who suddenly encounters a tragic accident or a violent incident. It's just one more way the hunter can be a more well-rounded person, better prepared for "life in the real world."

As Carlos Gallinger describes it in "Hunting in the Human Way":

> When a human being kills something, especially
> something large like a deer or a bear, it can cause a
> flood of emotions such as fear, pride, joy, guilt, and
> remorse.[140]

As Gallinger points out, ancient hunters had a culture that helped them understand and accept their feelings. The elders or the shaman would provide their wisdom. There would be rituals to guide the hunter through the experience. The culture would address the needs of the hunter's mind and spirit. He continues,

> In today's world a hunter often does not have access
> to this type of wisdom, let alone rituals. There may
> be some older hunter or guide in camp that know[s]
> some of these things, but due to the culture and cus-
> tom of our times, such things are not always talked
> about in a manner that they should be.[141]

At the moment of the kill, the hunter's focus is intense. For that moment in time, the predator and the prey are so tightly connected that it feels as though we are one. We are frozen in time; our souls are one. The prey will live on through the predator. We do not take it lightly. As is often said when a life is given so another may live, "Make it count. Make it matter."

And on those occasions when I choose to kill, to claim some small part of nature's bounty for my own, I do so by choice, quickly, with the learned efficiency of a skilled hunter.

—M. R. James, "Dealing with Death" in *A Hunter's Heart*

Some people may not understand the emotional response of big game hunters or plains game hunters, but they surely can understand the turmoil experienced by the dangerous game hunter. Dangerous game animals include both carnivores and herbivores, predators and prey. What makes them dangerous is that they will kill humans, often with very little provocation. They are dangerous to man, are killers of man, and, if provoked at all, will strike with all their might. There is no doubt that having your own life directly threatened changes the texture of the hunt, with prolonged adrenalin and heart rates approaching "condition black."[142] Indeed, this may well be where the hunt and the gunfight merge, and this will be the focus of the next section of this book.

In every kill, there is a spirituality that outsiders cannot fully understand. For example, Viv says in her account that when she saw the limping buck in the far tree line, she "knew this was the buck for me." She was then attached to that buck. As our Native American traditions say, the animal had come to her. She was now the one to put him out of his misery, to let him have an honorable, meaningful death. He would not fall exhausted in the forest and be torn apart by coyotes and ravens. He would die with a clean shot, his flesh would be appreciated by Viv's family, and his trophy would be given a place of honor.

When Linda came out to Viv's deer to help, she petted the buck, wiped the blood from his coat, and tucked his tongue into his mouth, and she thanked the deer for coming to Viv. Then she and Viv posed the buck for pictures. The pictures (as well as the antlers and the story that Viv wrote for this book) would become the lasting artifacts of this beautiful buck.

> Hunting stands out among sports because it requires extreme self-control, what with the adrenaline pumping, the lure of the prey, and the very small window of opportunity; it's too easy to pull the trigger or release the string prematurely. Often the best skill a hunter has is being able to hold the shot or forsake it altogether. It takes true discipline...excellence as hunters is the goal.
>
> —Jesus Ilundain-Agurruza, "Taking a Shot"
> in *Hunting—Philosophy for Everyone*

Campfire Story: Coming Home to the Kalahari[143]

When we won a hunting safari, one of our "someday" bucket-list dreams was suddenly promoted to reality. We immediately upgraded what we won to include two hunters and four trophy animals. We had nine months to prepare for the trip: guns, ammo, inoculations, flights, paperwork. Before we knew it, we were landing in Windhoek, Namibia.

Day One—Kalahari Heaven
Our PH, Volca Otto, picked us up at the airport and drove us to the Intu-Afrika Kalahari Private Game Reserve, our home for the next five days. The red sand dunes of the Kalahari Desert rise up and stretch for miles, with grassy meadows between each one. It is beautiful and an ideal place to "spot and stalk" for plains game. What we didn't know at that first moment was how we'd feel at the end.

Once the official welcome (with fresh juice in wine glasses) was complete, we headed out to confirm our rifle zeroes...This was also Volca's opportunity to assess our marksmanship and determine what kind of shot we'd be able to make in the field. Since there was still daylight, we started the hunt on twenty-five thousand acres of dunes and meadows and dry pans. We eventually spotted an oryx that looked promising, and we stalked (on foot) over several dunes before we acknowledged that he had outsmarted us.

We returned to our accommodation. It was a private cottage, walking distance to the main lodge, where we enjoyed excellent food (always at least one game meat on the dinner menu) and the company of our PH and his wife. From the dining area, in the gathering dusk, we watched animals come to the nearby watering hole…oryx, springbok, wildebeests, giraffes.

Zebra, Oryx, and "Old Blue"

The next morning, we came across the zebra stallion we were after, and after a brief stalk, Linda settled into a bipodded prone position on the top of a dune and made a single shot from 225 meters.

The following day was Keith's kudu hunt. We travelled to a nearby conservancy and spent the day searching for kudu. We found only females and young males, not harvestable. Toward the end of the day, we happened on a herd of eland. These are the biggest antelope in the world, and the one our PH had spotted was an "old blue."[144] We stalked up on the herd several times, but "old blue" wouldn't let us come closer than about four hundred or five hundred meters. Finally, in the gathering dusk, Keith shot him at 470 meters.

The next day, we were back on the beautiful Kalahari searching for oryx. We saw about fifteen hundred oryx before we found the beauty we had seen on the first day. Linda shot him from 245 meters, and her hunt was complete. As Linda lay on the warm red sand, giving her animal some private time…she was thinking that her Paleolithic ancestors must surely have hunted the ancestors of this magnificent oryx.

Kudu Eyes

Once again, the search for kudu was on. We travelled to a property near Windhoek, a rocky, mountainous area on the Khomas Highlands. We spent the day driving and walking and searching. Kudu are hard to spot, and they'll let you walk right by without alerting. We finally got our "kudu eyes" and started seeing them, but none of the males were big enough to take. We spent the night in Windhoek and had dinner at the must-do Joe's Beerhouse. We returned to the kudu property the next day, and just as we were driving out, one of Volca's trackers spied a warthog about three hundred meters away. Warthog was also on Keith's list. When Volca said, "Let's go take a look," we knew the hunt was on! After a short stalk through a stream bed and into the tall grass, Keith took the warthog at 130 meters.

Back at the lodge that evening, we were treated to a traditional and beautiful "dinner on the dunes." Firepit, white linen, champagne…the works! And our own game meat (eland and oryx) on the barbecue!

The Rest of Namibia

We spent the next five days touring, with two days to absorb the vast loveliness of the Namib Desert. From one of the driest places in the world, when you climb one of the highest sand dunes in the world, you can see the Atlantic Ocean. It is fascinating, spiritual, and surreal.

The last couple of days of our Namibian dream were spent along the Skeleton Coast, the Atlantic shore. We stayed in Swakopmund (a lovely German colonial town and beach resort), enjoyed seafood dinners at "The Tug" and "The Jetty" and a catamaran trip to the nearby Walvis Bay seal colony.

Our final night was in Windhoek. The car rental company drove us to the airport, and then we were gone. But we left our hearts in Namibia. We're already planning to return in a couple of years. That elusive kudu, those spiritual dunes, and our Paleolithic ancestors are drawing us back.

Chapter Ten—Summary

Modern perspectives and (all too often) cultural arrogance tend to distort and define how we look at our ancient ancestors. But if we examine the matter closely and impartially, it becomes clear that we truly hunt on the shoulders of giants. The hunting skills and techniques of our distant ancestors brought us to our current position at the top of the food chain. Indeed, they established the necessary foundation, the roots upon which our civilization has been built.

Although most of us no longer possess the hunting skills of our distant forebears, we are indebted to them, and they still have great influence upon who we are and what we do. Inside each of us are the same brain and endocrine system of those early hunters. To feel complete, we must see, acknowledge, and celebrate man the hunter. The skills needed to hunt are many and varied even today, placing demands on every aspect of the hunter's being. As we practice marksmanship, hone field craft, strengthen our bodies, develop hunter's eyes, and even learn to cook the meat resulting from the hunt, we realize our true kinship with our ancestors. We recognize and respect their lives and history as we fully utilize the skills of our heritage to become holistic people ourselves.

If we ignore this aspect of ourselves, it can place us at peril as we navigate the modern challenges of global ecology, conservation, wilderness preservation, and the well-being of humanity in the centuries to come.

In the next section, we'll look at how hunting skills and techniques—and thought processes—are related to law enforcement and warfare.

Section IV

It's What I Do

As we've said before, what drives humans are feeding, breeding, and security.

This section focuses on security: self-defense, law enforcement, and the military. The links between hunting and modern security may not be obvious at first. The roles of the professionals (law enforcement and military) can help to inform the hunting experience. And, importantly, the experience of hunting can help these professionals understand and accept their roles.

Chapter Eleven

On Killing

> As I said in *On Killing*, our objective must be:
> "no judgment, no condemnation,
> just the remarkable power of understanding."
>
> — Lt. Col. Dave Grossman, *On Combat*

Where does killing fit into the scheme of things? The person who is engaged in self-defense may inadvertently kill. The police officer who serves to protect his neighborhood may also kill. The soldier who may travel far to battle may be tasked to kill. Each of these has a different situation from the hunter. First and foremost, unlike the hunter, their adversaries are humans. Secondly, they will attempt many actions before they resort to killing…yes, even on the battlefield! And finally, they mostly view themselves not as aggressors but as defenders. They see themselves protecting the lives, or the way of life, of those who are close to them.[145]

As anthropologists have pointed out, it is a large step from ancient to modern killing. Our ancestors may have killed for the same reason lions kill. They needed food. They needed to ensure their DNA survived and thrived. Our biological imperatives would have driven our ancestors to kill.

In modern times, we have strong ethics against killing our fellow humans. Rule of law strictly controls individual aggression. (Or at least, that is the intent.) Only under direct threat of death does an individual allow himself to kill another. For the police, it is similar, but the key distinction is that police are upholding the rule of law and may not necessarily be under direct threat themselves. They are protecting a social institution and may incidentally be protecting an individual. In warfare, the individuals who are fighting against each other usually don't know each other at all. The warriors on both sides are protecting their side. This is socially sanctioned killing. As a friend of ours pointed out, we can choose to be peaceful, but none of us is harmless.

Having said that, the tools and skills required in hunting are useful to warriors and defenders. In fact, when Keith and Linda teach their courses, they often emphasize how valuable some hunting experience is for police and military personnel. And Dave, in his seminars for law enforcement and military audiences, tells his students that hunting is the best possible preparation for the combat situations they may face in their professions.

> Many veteran hunters, upon hearing accounts of nonfirers, might say, "Aha, buck fever," and they would be quite right.
>
> — Lt. Col. Dave Grossman, *On Killing*

Self-Defense

Self-defense is surely rooted in our enduring requirement to protect ourselves from predators, acquire food, secure the right to mate, or ensure some territorial (or asset) advantage. Indeed, modern criminals are often called *predators*, and we often describe them as "stalking their quarry" or "preying on their victim." The word *quarry* was originally from the Latin *cor* or *corata*, "the heart" and other entrails. The word *victim* comes from the Latin *victima*, "a person or animal killed as a sacrifice."

Military

The history of the military usually starts with the first recorded battle. This is the confrontation between Sumer (current Iraq) and Elam (current Iran) in about 2700 BC near modern Basra. The Hebrew Bible describes this battle. As time has gone on, warfare is often credited with improving the technology of killing. It is war that can both warrant and afford to invest in technology. The Prussian military analyst Carl von Clausewitz, in his book *On War*, calls it the "continuation of politics carried on by other means."[146]

War was distinctly different from hostilities exchanged in the hunter-gatherer world. In those days, individuals or families might have had disagreements that resulted in aggression, raids, and deaths. On the other hand, war is a creation of the state, where the fighters are not necessarily the offended parties (at least not directly).

Community Defense and Law Enforcement

Community defense (whether protecting family, tribe, or clan) probably arose when, like baboon males, the strongest were sent out to guard the edges of the group. Our very ancient ancestors, who lived in caves, likely used fire to "keep the wolf from the door" as well as for cooking and warmth.

Although legal codes go back at least to the time of Hammurabi (reigning King of Babylon from 1792 BC to 1750 BC), law enforcement as a specialty is a relatively modern invention. Prior to Hammurabi, "kin policing" or tribal policing was usually the norm.[147] The Roman Centurions were primarily soldiers, but they were used to enforce civil law. They were known to patrol markets and common areas, watching for lawbreakers.[148] After the fall of the Roman Empire, through the Middle Ages, many societies once again relied on the individuals to take responsibility for themselves, by the grace of the state.

Gradually, community policing became more formalized, and eventually, a "constable" headed a hierarchy of volunteers. The constables in an area would come under the authority of the "Shire Reeve," or sheriff. Near the end of the fourteenth century, the "justice of the peace" emerged as a separate role. They had the authority to act as judges (and as juries for civil matters). Centuries later, Sir Robert Peel established the Metropolitan Police Services in London, headquartered on a little side street known as Scotland Yard.

> Robert Peel's system was a success, and by the mid-19th century large American cities had created similar police forces. In London, the policemen were so identified with the politician who created them that they were referred to as "Peelers" or—more memorably—"Bobbies," after the popular nickname for Robert.[149]

> A nation that draws too broad a difference between its scholars and its warriors will have its thinking done by cowards, and its fighting done by fools.
>
> —(attributed to) Thucydides

Killing on Purpose

> In "primitive" societies…human beings are merely one's own little tribe; the rest [other tribes] are nonhuman, like animals. Killing animals is of course acclaimed, and nonhuman bipeds of the neighboring tribe are equally objects of prey.[150]

In ancient times, many different tribes lived on this earth concurrently. We don't know exactly what happened when they encountered each other. Based on our human genome, sometimes they interbred. But all these tribes are now extinct. We don't know why, exactly, they died out. The strongest and most dominant may have killed them off, treating them as we would any animal that threatened our security, food, or territory.

In *Blood Rites*, Barbara Ehrenreich advances the theory that we are hardwired by the events that our earliest ancestors endured. And they spent most of their lives in fear of other predators. As the balance changed, we were more often predators than prey. Ehrenreich says our rites and rituals often reflected the deep trauma of having been prey. And we were a special kind of prey. We had a large prefrontal cortex.[151]

In modern times, we see the interplay of prey and predator. We see the victim and the bully. We see the defender and the aggressor. We may see these features in ourselves. At times, we struggle to find a balance.

In the animal world, the best-known creatures who indulge in war are "army ants." The ants migrate across feeding areas as aggressive predatory foraging groups, killing every little edible in sight. If they encounter another group of ants, they systematically dismember them, leaving them to die in the desert sand. And if they can, they decapitate the enemy as they advance.

Genocide

In *The Third Chimpanzee*, Jared Diamond says, "Genocide has been part of our human…heritage for millions of years."[152] Yet we tend to think that genocide is unique to our era. Certainly, genocide numbers tend to be greater because our population densities are higher, our killing technology is better, and our communications are more effective. Diamond adds, "As the growth of world population sharpens conflicts between and within societies, humans will have more urge to kill each other, and more effective weapons with which to do it."[153]

It is worth noting that the last genocide on the North American continent was the genocide of Native American tribes, both literally and culturally. Jared Diamond reports that there are many quotes of famous Americans regarding Indian policies that are very clearly genocidal in intention. For example, President George Washington said, "The immediate objectives are the total destruction and devastation of their settlements." President James Monroe said, "The hunter or the savage state requires a greater extent of territory to sustain it, than is compatible with the progress and just claims of civilized life...and must yield to it."[154] Interestingly, this agricultural superiority citing "progress" is a pre-echo of the sentiments that today's modern hunters hear from the antihunting activists, with an underlying theme of "hate the hunter." In fact, the genocide of the native North Americans can be seen as not only about killing off the Indigenous people but also about killing off the hunter-gatherers.

> Because hunting is "a deep and permanent yearning in the human conditions," says Ortega, there is a chronic fury in all people to whom it is denied. When the uninitiated are allowed unexpectedly to hunt they trample its subtle rites, ignore its awe...and disregard...what is divine.
>
> —Paul Shepard, *The Tender Carnivore and the Sacred Game*

Murder

Most modern humans struggle with the idea of killing on purpose. Most moral codes and religions have strict prohibitions against killing other humans (although there are certainly many exceptions).[155] The Judeo-Christian traditions say, "Thou shalt not kill," and many Christians debate the difference between "kill" and "murder" in the Ten Commandments. Most modern translations cite this passage as, "Thou shalt not murder."[156]

No matter the religious or philosophical beliefs of an individual, killing to defend your own life is usually considered acceptable. So is killing at the request of the state to defend the lives of your nation.

As we've noted before, animal predators sometimes appear to kill for no apparent reason, at least not a reason that's apparent to us, strive as we might to rationalize it. As Jared Diamond writes in *The Third Chimpanzee*, studies have shown murder occurs in many animal species.[157] Indeed, Diamond cites our pedigree in xenophobia, massacres, and genocides. He notes that the chimpanzee is xenophobic and treats members of his own band differently from members of other bands:

> Common chimps…carried out planned killings, extermination of neighboring bands, wars of territorial conquest, and abduction of young nubile females. If chimps were given spears and some instruction in their use, their killings would undoubtedly begin to approach ours in efficiency.[158]

These animal examples notwithstanding, the modern cultural taboo of killing is strong in us. Our warriors often have a great deal of difficulty in killing another human being. Even though they've been trained. Even when they are under an active threat that's endangering their lives and the lives of their fellows. In *On Killing*, coauthor Dave writes extensively on this phenomenon:

> Why did these men fail to fire?…There is within most men an intense resistance to killing their fellow man. A resistance so strong that, in many circum-

stances, soldiers on the battlefield will die before they can overcome it.[159]

Historically, to counteract the killing taboo, the training and culture of both the police and the military emphasize that the enemy, the bad guys, are subhuman. The bad guys are pigs, rats, or swine. The good guys are our brothers. And like brothers, the good guys are "all for one, one for all." As long as the bad guys are subhuman, predatory animals, there's an element of purity in our war.

> "To me they were less than animals."
>
> —Vietnam vet, as quoted in *On Killing*

> "To me they were just groundhogs. I had killed lots of groundhogs as a kid growing up on the family farm."
>
> — Capt. Keith A. Cunningham, Vietnam vet

It's easier to justify and rationalize killing when we are defending the cause. In *The Victim Cult*, Mark Milke describes decades of Germans positioning themselves as victims of nations and ideologies prior to launching into war.[160] And as Diamond points out in *The Third Chimpanzee*, "Even Hitler claimed self-defense in starting World War II: he went to the trouble of faking a Polish attack on a German border post."[161]

Many wars, however, have a strong offensive aspect. These offensive exploits are usually marketed as preemptive: "If we don't head them off at the pass, we'll be slaughtered on the other side of the mountains." It may be true, but it doesn't matter whether it is or isn't. As long as the warring party believes they must defend themselves through a preemptive strike, they'll be able to rationalize their actions.

> Our first step in the study of killing is to understand the existence, extent, and nature of the average human being's resistance to killing his fellow human...Killing is a private, intimate occurrence of tremendous intensity, in which the destructive act becomes psychologically very much like the procreative act.
>
> — Lt. Col. Dave Grossman, *On Killing*

Killology

Coauthor Dave is a leading authority on killing, having written a book by that name and coining the term *killology* to describe the scientific study of killing.[162] He asserts that while killing to hunt is a normal part of our human ancestry, we now have a deep-seated cultural aversion to killing members of our own species. When this aversion might have started is open for debate, but Dave says:

> There is ample indication of the existence of the resistance to killing and that it appears to have existed at least since the black-powder era. This lack of enthusiasm for killing the enemy causes many soldiers to posture, submit, or flee, rather than fight; it represents a powerful psychological force on the battlefield; and it is a force that is discernible throughout the history of man. The application and understanding of this force can lend new insight to military history, the nature of war, and the nature of man.
>
> ...Most primitive tribes took names that translate as "man" or "human being," thereby automatically defining those outside of the tribe as simply another breed of animal to be hunted and killed.[163]

We do something similar, Dave goes on to point out, when we use slurs to refer to the enemy. Some of the tribes of Borneo (before contact and some to this day) live in such isolation that they still categorize any people outside their own tribe as "not human" and will kill any intruder without a second thought.

So, how is it that modern man comes to war with a resistance to killing? What has changed? One striking difference between ancient and modern conflicts is that ancient conflicts were immediate and elemental. They were about killing someone who was directly threatening you and your family. This is a hard reality. Killing someone who is simply (and probably temporarily) wearing the uniform of another country is completely abstract. This abstraction demands the warrior to believe in a cause. The ancient warrior would draw directly on his survival drive. The modern warrior often identifies strongly with warriordom. He respects the enemies for their skill, their courage, and their sacrifice.

Another difference between ancient and modern times is the background of the individuals who come to war. The hunter-gatherer faced killing daily to feed himself and his family. The early farmers faced killing seasonally, but eventually, the killing of livestock was moved away from the farm to the slaughterhouse. The city dweller hires out his killing and takes his meat from the grocery shelf. This gradual change has made modern people mostly unfamiliar with killing and dying. Importantly, they lack the inoculation their forebears received.

A final difference between ancient and modern warfare is the way men are inducted into the military. Conscription of one form or another has been common throughout history and often at the pointy end of a sword. Modern armies are mostly populated by volunteers, though conscription was used by Canada and others for the World Wars and by the United States as recently as the Vietnam War. But the bulk of the modern militaries are made up of free men (from cultures that increasingly emphasize the value of life), and free men can afford to have their own opinions about killing. In modern wars, the average soldier's taste for killing has been diminished. When examining modern warfare, historian Gwynne Dyer believes "the vast majority of men are not born killers."[164] Research bears this out.

Wild animals rarely engage in what we would call "war." But there are examples of what we might call "guerrilla war," or raiding. Animals must be willing to fight to the death for territory. And they often fight for the right to breed. Animals sometimes can display their fangs, antlers, or horns (or they can use their trumpet or roar) to intimidate without having to engage. But in modern war, generals do the posturing, and the infantrymen provide the real deal. Having soldiers not firing or not aiming to kill is neither appropriate nor effective. In *On Killing*, Dave explains:

> To address this deficit in hit ratio, the modern militaries changed their training programs after WW2. Marshall's contributions to the U.S. Army's training program increased the firing rate of the individual infantryman from 15 to 20 percent in World War II to 55 percent in Korea and nearly 90 to 95 percent in Vietnam.[165]

> The effectiveness of modern conditioning techniques that enable killing in combat is irrefutable, and their impact on the modern battlefield is enormous.[166]

When Keith was going through his Canadian Army sniper course, conducted by the Second Battalion of The Royal Canadian Regiment,[167] the head instructor, a seasoned and clever teacher, once told the class to "aim for the button." What he meant by this was to pick a small part of the target and aim specifically for it. A slight miss would still result in a lethal hit. If you were aiming for the entire area, like the whole chest, then a slight miss might result in a nonfatal hit or a complete miss. On the final exercise, each of the students received orders to travel cross country, in enemy territory, locate a specific target, and kill it. The targets were mannequins, dressed up like generals. After locating the right one, Keith made his shot and still has the button with a neat bullet hole through it.

> To aid in overcoming this resistance [to pull the trigger] it is helpful if you can will yourself to think of your opponent as a mere target and not as a human being. In this connection you should go further and pick a spot on the target. This will allow better concentration and further remove the human element from your thinking.
>
> —Bill Jordan, as quoted in *On Killing*

But Real-life Targets Don't Come with Sticky Dots!

One day, we were watching a TV show hosted by an archery hunter. He was using a life-sized deer target with a faint line showing the lethal zone. He commented that many archers were content if they could just hit the lethal zone anywhere, and he pointed out that they often had not zeroed correctly. Then the archery hunter reached off camera and brought into view a sticky dot, which he placed in the middle of the lethal zone. He said that this would help the archer to identify the correct point of impact. And if he is not hitting it, then maybe his zero is off.

Many of his students were quick to say that their real-life targets don't come with sticky dots.

But his real-life targets did, the hunter replied. He had shot at the dot so many times that when a real deer showed up, he could see the dot in his mind's eye. He could quickly locate the right place to aim. And with the bow zeroed properly, he could guarantee an ethical hit. Even if he missed the dot slightly, he would still have a good hit.

As coauthors Keith and Linda say, "Aim small—hit small."

The Natural Soldier

Is there such a thing as a "natural warrior" who can kill ethically and without remorse? Gwynne Dyer says:

> There is such a thing as a "natural soldier": the kind who derives his greatest satisfaction from male companionship, from excitement, and from the conquering of physical obstacles. He doesn't want to kill people as such, but he will have no objections if it occurs within a moral framework that gives him justification—like war—and if it is the price of gaining admission to the kind of environment he craves.[168]

Coauthor Dave emphasizes:

> There is 2 percent of the male population that, if pushed or if given a legitimate reason, will kill without regret or remorse. What these individuals represent—and this is a terribly important point that I must emphasize—is the capacity for the levelheaded participation in combat that we as a society glorify and that Hollywood would have us believe that all soldiers possess.[169]

With a couple of modifications, this description of the "natural warrior" could be a perfect description of the "natural hunter." The natural hunter derives his greatest satisfaction from the solitude of the wilderness, from the excitement of the stalk, and from conquering the hardships of terrain and weather. He doesn't want to kill animals as such, but he understands it is the consummation of the hunt. There is a percentage of the population who, when given a legitimate, ethical reason, will kill an animal without regret or remorse. What these individuals represent is the natural participation in a hunt that echoes through the long history of our ancestors.

And thus huntology (the study of hunting) also looks at the role of humans killing their fellow humans. Our study here can contribute to our understanding of killing in combat, and the study of killing in combat can contribute to our understanding of hunting.

> Combat is of such emotional intensity that strange elations can come from the act of killing.
>
> —Colonel James R. McDonough, as quoted in *On Combat*

Feeding, Breeding, and Security: All's Fair in Love and War

Our modern world tends to be very analytical. We tend to think of survival as an individual thing. We tend to think that saving one life is the only way to save all lives. Mother Nature is, on the other hand, naturally holistic. We can better understand her by looking at all the parts and especially the connections among the parts.

Our brain is built with survival in mind, and all the functions and brain chemicals organized for survival are co-located and largely overlapped. Our "animal" brain really doesn't distinguish much between feeding and breeding. It doesn't distinguish much between the exhilaration of the successful hunt and the exhilaration of successful mating. During both acts, the same portions of the midbrain are activated, and the same brain chemicals are released. This portion of our brain likes survival and celebrates with exhilaration, whether we are feeding or breeding. That's the holistic system that has made *Homo sapiens* so remarkably successful on this planet.

> The midbrain has no philosophy, no hesitation, and no regret…It acts decisively…It never apologizes, never looks back, and sheds no tears.
>
> — Lt. Col. Dave Grossman (quoting a correspondent), *On Combat*

Similarly, the ancient animal brain doesn't distinguish much between defense and offense. Again, they are the yin and yang of a specific aspect of survival…security. Our animal brain doesn't know the difference between attacking to get food or resisting an attack to keep from being food. The same portions of the midbrain are activated and the same brain chemicals released during fight and flight. The ancient brain likes security and ensures it with a dump of the same chemicals whether we are in fight or flight mode.

What's interesting is that the ancient brain responds much the same way no matter which of the "four F's" we're engaged in. Ah, yes…the four F's: fight, flight, feeding, and…breeding. And when we are doing any one of those, the brain's chemical pathways are reinforced for them all.

Target Reaction

The hunter knows about killing and about dying. He has seen animals die by his own hand. But hunters are by far the minority of modern humans, most of whom have never seen the death of anything bigger than a housefly. Because of this lack of familiarity with killing, most modern people don't have any idea as to how an animal reacts when it is killed.

Police students have asked Keith and Linda to include a module in their courses about putting down an animal that's been injured in a road accident. Today's modern police often have no experience with hunting, no experience with injured animals, and often no experience with killing. One cop told us that dispatch had sent him to deal with an injured deer on the roadside. The cottager who'd been involved in the accident was screaming at the cop, "Kill it! Kill it! It's still alive! It's in pain!" The deer was, in fact, already dead. The cottager mistook the final twitching of the leg muscles as a sign of life when, in fact, it was a sign of death. Experts tell us that when it is alive, the brain keeps the animal from twitching. When an animal is brain dead, it can no longer control those muscles. All that's left is latent signals from the primitive nervous system.

There are many different things that can happen when an animal is killed. Years ago, we used to teach our police sniper courses on military bases. One such base had a range that was nine hundred meters deep and about fifteen hundred meters wide, over three hundred acres of lovely, grassy, groundhog heaven. Range control considered the groundhogs to be vermin because of the holes that they put in the range floor. These holes endangered our soldiers during rundowns (fire and movement exercises) on the range. So, we would take our sniper students out on the range in the evening and have them practice their skills on the groundhogs. It was a great education for them, not only developing their skills at setting elevations, judging wind, and calling fall of shot (watching swirl and splash) but, more importantly, also learning what happens when a living creature is shot. They learned that there are many different target reactions, and they were being inoculated… learning how to deal with their own feelings about killing.

Cops and soldiers who have never been hunting usually don't have killing experience until their first lethal encounter on the job. They don't know what to expect because they haven't been inoculated. They are at increased risk of being traumatized and later suffering post-traumatic stress. Coauthor Dave says that our groundhog shooters, through this inoculation process, have significantly reduced the likelihood of a post-traumatic response should they have to kill a human being in the years to come.

Campfire Story: Linda's First Deer

The light was fading when I shot the deer. He jumped and ran to the nearby tree line. I called Keith on the radio and told him I had a buck down. Although the deer had only gone about fifty meters, it was dark by the time we tracked him down. We found him nestled under a juniper bush. You could imagine him saying to himself, *Oh, I'm just so weary! That looks like a nice dry bed. I think I'll just stop a while and rest.* He had taken his last breath there in the forest.

We pulled him out to the clearing so we could get better light on him. It was a frosty night, and the little snow covering helped him glide along the ground. We posed him for a picture, making sure he looked regal and peaceful…after all, this would be how we would remember him. Then we stretched him out on his back, and Keith began to field dress him. As the knife slid up his belly, a white mist rose from him. It was the heat of his body, condensing in the cold night air. It was his animal spirit, leaving on his journey to join those who have gone before.

History of the Spirit

For thousands of years, our ancestors may have respected the spirit of the body and respected the body as the last container of the spirit. It doesn't matter if you believe the spirit is religious (the soul) or if you believe the spirit is the life force (the mind) or if you believe the spirit is purely biological (the last breath), it's clear there is a difference between a live animal and a dead one.

The consequence of hunting is all reward: food on the table, pride in the achievement, hide and horns in the trophy room. Inexperienced hunters may not be able to eat the meat right away and may have to wait several days before the image of the live animal fades. Over time, over many hunts, the hunter gets somewhat inoculated, but most hunters still have a ritual; most have a little mourning for the lost life. It becomes part of the hunting process, one they expect and one they pass through.

Both law enforcement and military personnel have benefited from "returning" to the hunt even if they've never hunted before. It seems that, for at least some individuals, you can retrofit your brain connections during the hunt and repair some of the trauma. Many hunting and veterans' groups are now organizing hunts for both able-bodied and disabled veterans.

Our knowledge of hunting can help us understand post-traumatic stress (PTS), and our study of PTS can help us understand hunting. Additionally (and this is a key point), hunting can be a powerful tool for helping our veterans transition from PTS to post-traumatic growth.[170]

Chapter Eleven—Summary

There is much to learn at the intersection of hunting and combat. The study of killing in the realm of hunting has a great deal to contribute to understanding and scholarship within the domain of human behavior.

The "natural warrior" has much in common with the "natural hunter." Indeed, the realm of hunting may fulfill a natural "need" that appears to exist in much of humanity. Killing in combat can be viewed as a pathology or, at best, a tragic necessity: something we hope that humankind will outgrow and leave behind. But killing in hunting can be viewed as a healthy and natural part of the circle of life. Hunting teaches us about killing and about death and therefore can be a valuable preparation for those who need to be familiar with those realities in their line of work.

The natural hunter derives enormous satisfaction from the fellowship of other hunters and from the solitude of the wilderness. Such individuals rejoice in the excitement of the stalk and the challenge of the hardship of terrain and weather.

To those who follow the hunter's path, the act of killing game is just one part of a much larger and profoundly fulfilling endeavor. These individuals have found nourishment (both emotionally and physically) in the deep roots of our most ancient heritage. And we believe that such intense satisfaction is available to others who would choose to hunt and listen closely to the echoes of our ancestors.

Chapter Twelve

On Combat

Hunters get an adrenaline dump before (or during) the firing of a shot. And then they have the field work to help use up the adrenaline and rebalance the nervous system. Soldiers get the adrenaline dump and then have a battle to eat up the adrenaline. Cops get the adrenaline dump and then maybe a couple of shots and then maybe nothing…and they must go to the gym, on a long, hard run, or perhaps have some intense sex to wear off the adrenaline. All of this adrenaline comes from our primitive brain. Its design keeps us safe and enables us to hunt dangerous animals.

Certainly, you can get an adrenaline dump from anything that makes you anxious: public speaking, exams, an employee review, or a disciplinary meeting. And of course, you can get an adrenaline dump from a potentially dangerous situation, like getting lost in the bush, a traffic accident, or a natural disaster. Many people get adrenaline dumps from phobias (irrational uncontrollable fears) about many things that don't really present an immediate danger—spiders, snakes, heights, darkness. Snakes are a very common phobia, but not everyone is afraid of snakes, in fact, only about 15 percent of the population. Most people are afraid of confrontation, especially physical confrontation. Experts call it the "Universal Human Phobia." It affects about 98 percent of the population. It is the irrational, uncontrollable fear of interpersonal human aggression.

Most of us don't want to live our lives expecting violence at every turn. So most of us go around in "condition white," a resting state, or "condition yellow," an attentive but relaxed state. Trained warriors and hunters are usually in "condition yellow" when they are out in the world. If something happens that demands their specific attention, they direct their focus on an immediate potential threat and then may escalate to a more intense "condition yellow" or "condition red," which is a peak-performance state with high energy but possibly some loss of fine-motor control.[171] If they top out beyond that (more adrenalin, higher heart rate), they can end up in "condition black," an over-aroused, low-performance state that compromises both judgment and fine motor control.

Hunters who are attentively watching their area are in "condition yellow." If a nice buck enters the clearing, they will quickly escalate to a higher, more focused level of "condition yellow," still maintaining fine motor control. Here, they'll be thinking about distances and angles, figuring out when the best shot might materialize. If it's a *huge* buck and they don't have a lot of mental training, they can easily slip into a state that's known as "buck fever." In a buck-fever state, they'll be overexcited. They may move into "condition red," which means possibly losing fine motor control and being unable to take the precise shot. They may be on the edges of panic or even into "condition black." In this state, hunters may forget to take the safety off and not know what's wrong with their rifle. They may cycle the action and dump live rounds on the ground without ever touching the trigger. They may look over the sight and shoot low, or they may freeze and not shoot at all.

> You do not rise to the occasion in combat;
> you sink to the level of your training.
>
> — Lt. Col. Dave Grossman, *On Combat*

Although it may be hard to see, the body's plan here is to make you more effective. In ancient times, that adrenaline would have helped the hunter. It would have improved his fight-or-flight response. It would have helped him to overpower or escape dangerous game. However, that reaction of the body is overkill (so to speak) when all you must do is stay still and pull a trigger.

There are a couple of other effects that accompany the adrenalin. One is that your blood drains from your extremities. This may reduce your fine motor control. It may make your face pale. The blood is being redirected to your large muscles, preparing your body for flight-or-flight mode. This may limit blood loss in a combat situation. And it may have contributed to the survival of our hunting ancestors in an encounter with fangs and claws.

Another effect that accompanies a high-stress situation is sensory distortion. You may have heard people say that "time stood still" or that it was like a "slow-motion video." Hunters, cops, and soldiers often report that they did not hear shots fired. It appears that the brain is screening sensory input. The brain focuses on what we need and tunes out all the rest. For example, if a lion roared as he leapt on you, you might feel the impact but not hear the roar. Soldiers tell of hearing an RPG (Rocket Propelled Grenade) coming but not hearing its impact. This internal defense may help us avoid sensory overload and confusion. Cops frequently report having "tunnel vision." This is so common that their training now includes "scanning" to counteract their tunnel vision. (It also includes side-stepping to take advantage of the bad guy's tunnel vision.)[172]

Humans have been both predators and prey. Alterations to our sensory perceptions may help in both situations. In *On Combat*, coauthor Dave writes:

> [These distortions]...may imply some kind of predator response, where the hunter focuses so exclusively on the prey, the sounds and motions of the rest of the herd do not distract it...[I]ncreased visual acuity would be useful, since fleeing prey needs to see every possible option and avenue of escape while staying alert to any ambushing or intercepting predators. Meanwhile, the pursuing predator fears nothing except losing the prey. So for it, tunnel vision is potentially a survival trait.[173]

Understanding these physiological processes, gained from the study of combat, can be of great value to the hunter. This information can forewarn and forearm the hunter so he is not "blindsided." When he must deal with altered perceptions, he is prepared. He better understands how his body is responding. He can appreciate that his ancient brain is simply trying to improve his likelihood of survival.

"Group-ness" and Bonding

Group bonding is an important step in the development of the most successful species in the history of this planet. In *The Righteous Mind*, Haidt asserts:

> [W]hen early humans began to share intentions, their ability to hunt, gather, raise children, and raid

their neighbors increased exponentially...When everyone in a group began to share a common understanding of how things were supposed to be done, and then felt a flash of negativity when any individual violated those expectations, the first moral matrix was born. The moral matrix is the glue that holds groups (families, tribes, nations) together...Victory went to the most cohesive groups...Natural selection favored increasing levels of..."group-mindedness."[174]

This groupness included the natural selection of individuals who were inclined to be good group members. This enabled the "strength of many" in obvious areas like hunting, defense, and war.[175] But it also helped us develop shared intentions, culture, and language. The modern hunt camp is likely a vestige of this "group bonding," with its underpinning in tribal behaviors. For modern warriors, group bonding is a reality. It starts with drill. Barbara Ehrenreich writes:

Historian William H. McNeill [suggests]...group dancing and military drilling...can be traced to the primordial sociality of the [ancient human] band confronting a wild animal.[176]

This relationship between group dancing and military drilling is probably easiest for us to visualize in North American native dancing, which was used for many social purposes, including war. The war dance would typically occur on the evening before an attack was to be launched. The dance, highly rhythmic and done in unison, was intended to give the participants a sense of community and purpose. A war dance often involved singing or chanting. The Māori in New Zealand perform the *haka* to display pride, strength, and unity. (You may have seen the All Blacks rugby team perform the *haka* as a pre-match ritual.)

Modern armies usually use chanting for marching and running. The importance of movement in unison is fundamental to military drill. Most soldiers have experienced marching while using "jody calls." This is a call-and-response format where one soldier initiates a line and the remaining soldiers complete it. This instills teamwork and camaraderie. The cadences may move to the beat of the quick-time march, the double-time march, or jogging. This serves the purpose of keeping soldiers "dressed," moving in step as a unit and in formation while maintaining the correct beat or cadence. In *The Righteous Mind*, Jonathan Haidt quotes World War II veteran (and distinguished historian) William McNeill:

> Words are inadequate to describe the emotion aroused by the prolonged movement in unison that drilling involved. A sense of pervasive well-being is what I recall; more specifically, a strange sense of personal enlargement; a sort of swelling out, becoming bigger than life, thanks to participation in collective ritual.[177]

Only a warrior can testify to the bond he shares with his brothers. As Haidt says, "Basic training bonds soldiers to each other, not to the drill sergeant."[178] The strength of this bond is indisputable. Accounts from all sources confirm that a warrior will give his life to save his brother warrior.

Hunting Skills Applied

The methods and tools used for hunting can also be used for self-defense. In a fight to the death, it is impossible to distinguish the predator from the prey or the aggressor from the defender. The animal kingdom illustrates this well. For example, kangaroos are prey animals. They are adept swimmers and will flee into waterways if that option is available. If pursued into the water, a kangaroo may use its forepaws to hold the predator underwater to drown it. Correspondingly, the crocodile (a predator) is well known for drowning its prey.

On land, the kangaroo will defend by holding the attacker with the forepaws and disemboweling it with the hind legs. This is what the big cats do to subdue their prey. In other words, when threatened, the kangaroo defends vigorously, using the same tools that predators use to kill.

Indeed, the deep roots of war reach back to our early conflicts with animals. In *Blood Rites*, Barbara Ehrenreich makes that point. She says that not only were hunting tools later used in war but so were hunting tactics like the wedge and the frontal attack. "Well into the modern era, in fact, light regiments, which advanced in a line much as Paleolithic hunters may have done in their drives against animals, were called 'hunters'—*jaegers, chasseurs*—and were dressed in green."[179]

The skills and the tools of warriors and hunters are similar. As we explore this in the following pages, we focus more on skills than tools. As coauthor Dave says,

> In the end, it is not about the hardware, it is about the "software." Amateurs talk about hardware (equipment), professionals talk about software (training and mental readiness).[180]

Marksmanship

> But never assume that simply having a gun makes you a marksman. You are no more armed because you are wearing a pistol than you are a musician because you own a guitar.
>
> —Jeff Cooper, *Principles of Personal Defense*

Keith was a member of Ranger Recon Team "Miami" while in Vietnam. He did a total of thirty-seven missions lasting from six hours to thirty days. From this combat experience, he concluded two things. First, in the end, it's luck that gets you through. You can improve on your luck with training, practice, attention to detail, alertness, and situational awareness, but in the end, it all boils down to luck. On at least a couple of occasions, he survived because the other side was a bad shot, and he was lucky for that.

The second point is that you can "never be too good a shot." He saw lots of soldiers shooting from the hip, sprayin' and prayin', with the rifle on "auto-get'em," feeling secure behind a wall of bullets and then panicking when they ran low on ammo. In Vietnam, it took approximately fifty thousand rounds fired for every hit achieved.[181] Can you imagine what it would have been like if there had been a hit with every two rounds fired, or even every five rounds fired; okay, let's get crazy—a hit for every ten rounds fired? The battle would have been over in half the time with fewer casualties on the good-guy side. You can never be too good a shot.

Coauthor Dave tells his police audiences:

We can't help growing old, and some of us can't help
putting on some weight as the years go by, but we
can still be one hell of a shot. There is an excuse for
growing old, and maybe putting on some pounds, but
there is no excuse for not being a damned good shot!

We once heard a story about a police officer who was advocating the carrying of extra magazines. It seems he was in a gunfight where he and a bad guy exchanged shots as they ran around a couple of parked cars shooting at each other. When the police officer was on his third magazine, he realized he had to get this over with before the bad guy got lucky with a shot. He remembered his training and the need for a sight picture for accurate firing. He then accomplished his goal with one well-placed shot. The idea here is to accomplish this well-placed shot within the first few shots fired and not wait until the third magazine. Luck is fickle and cannot be trusted—she is not always on your side. You can make her smile in your direction if you have trained and practiced properly. A sports reporter once commented to golfer Arnold Palmer about how lucky he had been lately with his putting. Arnold replied, "Funny thing about that; the more I practice the luckier I get."

> The great body of our citizens shoot less as time goes on. We should encourage rifle practice among schoolboys, and indeed among all classes, as well as in the military services by every means in our power. Thus, and not otherwise, may we be able to assist in preserving peace in the world. The first step—in the direction of preparation to avert war if possible, and to be fit for war if it should come—is to teach men to shoot!
>
> —Theodore Roosevelt

Field Craft

Many of the missions that Keith was on in Vietnam were uneventful—a "dry hole" as they say. The primary job of these Ranger Recon missions was to go quietly into a certain area and report on enemy activity. As Keith tells it:

> No enemy activity, a "negative sit rep," and it felt
> like being on a camping trip (almost). You got to
> sleep out under the stars. You got to look at the neat
> terrain (in the foothills one day, in the rice paddies
> the next). You got to take time to actually look at the
> vegetation and see how things grow in the jungles.
> You were able to take time to warm up your LRP

(Long-Range Patrol) rations. It was almost like being on a camping trip.

The big difference is that you still had to apply field craft. You put fresh cam paint on your face every few hours. You walked quietly. You spoke no louder than a whisper. You had to look everywhere at once. And you always, always kept your rifle within arm's reach. Just like hunting.

An avid hunter before going to Vietnam, Keith found it easier than most to develop the "hunting eyes." While hunting the snowshoe hare back home, he looked for the little mound of snow with a black spot at one end. That gave away the hare sitting motionless, relying on his camouflage to keep him out of sight. During the deer season, it was the straight horizontal line of a buck's back standing still among the vertical lines of the saplings around him. Sometimes it was the white patch on his throat or the smooth roundness of his ears. The buck made finding him easier if he nervously flicked that ear or his tail.

Recently we were hunting red deer stags. At first, they were hard to spot. But after a couple of days, we could see them bedded down in the bush. We learned to see the merest glimpse of their unique antlers, shaped unlike the shrubbery around them.

In Vietnam, the same principles applied. Keith tells the following story about urban police observation:

Years ago, I was teaching a police sniper course at a local agency. I was covering observation techniques and how to use binoculars for the first scan to look for suspicious areas, and then how to use a spotting scope for a detailed look. Suddenly, the sergeant came into the room. Clearly with an element of urgency, he required the students to go on a call. It seems that a bad guy had opened fire on a patrolman during a routine traffic stop and had run into a housing complex that was partially under construction. My students all instantly disappeared, one of them grabbing the spotting scope I had been using as a training aid.

The next day I got the rest of the story. After securing the housing complex, the sergeant decided that the team would conduct a house-to-house search. One of the students was scanning the area with his binoculars when he noticed one house had a broken window. He focused his spotting scope on the broken window and could see a red stain that looked suspiciously like blood. He drew this to the attention of the sergeant, and the sergeant decided that it would be the first house searched. It wasn't long before they had the bad guy in custody, having found him hiding in a closet. The moral of the story? At every opportunity, stay alert, be situationally aware, and develop and practice your "hunting eyes."

In our police sniper field craft, we use a series of progressively more difficult observation exercises to train the students. When they first start these exercises, they can only find the most obvious of articles placed within a relatively small area of observation. Gradually, they become sensitized. They learn to look for a glint, a dark spot, a hard or straight line, a shadow. They start to see anything that might indicate an object that's out of place. It takes a degree of conscious attentiveness, combined with "letting" the brain see something that doesn't fit the pattern. And everyone gets better with practice.

Physical Fitness

To be successful in combat, you need to have good overall physical fitness, including aerobic fitness. It's hard to be in control of yourself or the situation if you feel light-headed after having run up a few flights of stairs. You also need to have upper body strength. You need to have the endurance to be able to handle, use, and wear the required equipment for as long as needed. You need to be able to look at yourself and be confident that you're the officer you'd pick for a partner.

Of course, we include soldiers in this as well. The advantage soldiers have is that they train continually until it's time to go to war, and with all this training, they have ample opportunity to stay fit. Police officers, on the other hand, do a course, get issued their badge, and are at war (so to speak) for the rest of their careers. Each officer must plan fitness into his day and must have the willpower to make it happen.

Mental Toughness

> Courage is being scared to death
> but saddling up anyway.
>
> —John Wayne

Along with physical toughness comes mental toughness. It's the confidence you have knowing you can handle the worst-case situations, at least most of them. This allows you to stay in mental control.

Mental toughness gives you the self-discipline to stay alert and recognize the situation as it's unfolding. It allows you to be as alert at the end of the shift as you were at the beginning. For the soldier, mental endurance is about the need to stay as alert coming back from the patrol as going out. Mental toughness will keep you situationally aware—the need to stay in tune with your surroundings. Situational awareness is perhaps the most important mental skill, if you had to choose one as more important than another. Situational awareness can keep you alive, and the lack of it can most certainly get you killed. And part of this is having the mental discipline to be at the right level of awareness (or arousal), always in yellow and ready to go to more intense yellow or red as the situation develops. This same mental toughness keeps you from going to condition black.

Mental toughness means that when you are afraid, you're still able to do the right thing. Overcoming fear allows you to do your job, to run toward the gunfire while everyone else is running away. Even though your life could be over in the next few minutes, you overcome the fear by focusing on the things you need to do. There are many things you must do right; focus on them. Have "grace under fire."

US Army Mental Preparation

When we first saw material from the US Army on the mental side of combat, we were surprised. Ground-pounders being taught how to think about combat and what to think to get the best performance—it's straight out of sports psychology, and it is immensely powerful. The US Army program teaches what they call "battle-focused mental readiness." It emphasizes that warriors need to think, focus, and act in a positive, decisive way. They are expected to learn new skills, practice those skills, and perform those skills on demand and under stress.

On Combat and On Killing

In Keith and Linda's mental marksmanship seminars, we always tell our audience that *On Combat* is a must-read for brothers-in-arms.[182] In it, Dave Grossman advocates that you develop a "bulletproof mind" by being prepared, specifically through stress inoculation. Of course, the "bulletproof mind" is about mental toughness. You must be prepared to take action. You must use stress inoculation to prepare. Afterward, you must focus on "post-traumatic growth."[183]

One of the stories Dave tells in his Bulletproof Mind seminar is about Jennifer Fulford, a deputy sheriff in Orange County, Florida. She responded to a 911 call for a robbery, and it turned out that three children were trapped in a van parked in the adjoining garage. Fulford immediately said, "I'm going to try to get the kids." As she crouched beside the van, she reported hearing shots fired. Seconds afterward, she received and returned fire, killing the bad guy. She exchanged gunfire with a second robber and took several hits, including one that put her right hand and arm out of action. She dropped her gun but managed to grab it with her left hand and continued to return fire. The second bad guy fled the garage and was shot by one of the other responding officers.

The officer called to Fulford and asked her if she was okay, and she says in her presentations, "I'm like, 'No! Get me out of here!'"

Jennifer was shot seven times. Three additional bullets slammed into her body but were stopped when they hit body armor or equipment, yet she never stopped fighting. She credits her training sergeant with not only putting her through support-hand drills, including reloads, but also insisting that she stay in the fight.

Reflecting on her ordeal, she said, "Mindset is really important. I was not going to die in that garage. I am the product of my training."

Dave's book *On Killing* is another must-read for brothers-in-arms. As we said before, Dave deals with the psychology of killing in a way that no other author has covered the subject. He says, "Do not flee from your fears, master them."[184] His emphasis is on being able to accept the responsibility to make sure the good guys win. And if you're not willing to kill, he reminds you that the bad guys certainly are. He says that courage isn't being fearless;

courage is controlling your fear. He stresses that you need the same courage after the battle.[185] And he, like we do, believes that hunting inoculates you. It's a positive exposure to killing.

Secrets of Mental Marksmanship

When Keith and Linda wrote the *Secrets of Mental Marksmanship*, they were not only trying to describe the secrets but also to identify the similarities of their application to police officers, soldiers, hunters, and competition shooters. There are three key points in the book:

- By training your mental skills, you can control your ability to "get in the zone." This is where you perform at your best, on demand.
- You need to focus on performance, use positive reinforcement, and surround yourself with like-minded people.
- You need to persevere.

The tools are there; you just have to pick them up and use them.[186]

Principles of Personal Defense

No section on mental toughness would be complete without a tribute to Lieutenant Colonel Jeff Cooper, who believed that the primary tool for personal defense is the combat mindset. Cooper popularized the earlier-mentioned color code system as a method of describing mental arousal and readiness to deal with an attack. Cooper believed strongly in the importance of mental control. He said that "man fights with his mind. His hands and his weapons are simply extensions of his will."[187]

When it comes to aggressiveness in combat, Cooper turned to the animal kingdom for examples, including this one: a forty-pound wolverine drives a whole wolf pack away from a kill that the wolves worked hours to bring down. "Aggressiveness carries with it an incalculable *moral* edge in any combat, offensive or defensive."[188] And as do all three authors of this *On Hunting* book, Cooper advocates both hunting and competition to improve your ability to stay cool in a gunfight. About hunting he says:

> "Buck fever" is a classic affliction, and a man who has conquered it can be guaranteed to shoot carefully under pressure. While it is true that a deer is not shooting back, this is less significant than might at first appear. The deer is about to vanish, and, odd as it seems, fear of sporting failure is usually greater than the fear of death. This startling point is easy to prove. The average competitive pistol shot works and trains far harder to earn a little brass cup than the average policeman works and trains to acquire a skill that can save his life.[189]

The Combat Hunter

> Let him who desires peace
> prepare for war.
>
> —Flavius Vegetius Renatus, *De Re Militari*

The book *Three Day Road* tells the poignant story of two young Cree snipers fighting in World War I. At a critical point of the story, these experienced bushmen realize that they've been behaving like rabbits, and if they are to survive, they must behave like wolves. If they act like prey, they will be preyed upon. If they act like predators, they are much more likely to survive the war and return home.[190]

In Vietnam, Keith one day realized that he was tired of being scared, scared of dying. That was a turning point. From that day forward, he wasn't scared. He just focused on doing his job. He went out on his missions with a purpose. He took care of his men. He stayed alert. After about a month, he realized he was starting to believe that he'd be okay, that he'd be able to get through this thing called war, and he'd be able to return home.

The USMC Combat Hunter Program

In 2006, Gen. James Mattis asked the Marine Corps to develop a program

> to instill a hunter-like mindset in Marines, train Marines for increased situational awareness, proactively seek threats, and have a bias for action. Mattis wanted Marines to be the predators, not the prey.[191]

This request ultimately resulted in the Combat Hunter Program, which is described in the book *Left of Bang*.[192] The program emphasizes using intuitive decision-making, which is needed in uncertain situations where speed is critical. This type of rapid processing is based in pattern recognition, a skill honed by hunters. Key to intuitive decision-making is alertness and having the "offensive mind of a hunter instead of the defensive mind of prey."[193]

Ivan Carter, the professional hunter we've mentioned before, developed the observation portion of the program, which focuses not just on looking but also on truly seeing. The trackers in Africa are well-known for their vision and especially well-known for their ability to "see." The relationship between hunters and Bushmen and the military is nothing new; they have a history. As early as 1685, the Cape Dutch recognized the local aboriginals' skills as hunters of game and gatherers of natural foods. The Dutch referred to them as "Bosjemans," and thereafter, the natives were known as the Bushmen of Southern Africa.

The Combat Hunter mentality is very much like the successful game hunter mentality. As the authors say in *Left of Bang*:

Humans…make quick decisions, often based on patterns we have observed and learned: "If facing someone or something in the forest, our ancestors needed to respond instantly: friend or foe? Those who could read an expression in a flash more often lived to leave descendants, including us." Intuition is best used when a person has significant experience and knowledge, which guides that person's subconscious thought processes.[194]

Sniper Meditations

Of all the police and military roles, the role of the sniper is most closely related to the role of the hunter who engages in hunting from a stand or the spot-and-stalk style of hunting. In Keith and Linda's Police Sniper courses, they emphasize some of the attitudes and personality traits of an effective sniper. A portion of this course goes as follows:

> Not everyone is suited for this important tactical position. The entry teams, pumped and motivated by being part of a team about to engage in a sudden life-and-death situation, will kill in a react-mode manner. The sniper is often out of immediate harm's way and, when required, must kill deliberately and most intentionally. He is often the last resource called upon and is expected not only to incapacitate but also to most certainly eradicate his target.
>
> He is called on because the target area is small, often partially hidden behind innocent hostages, or he must shoot among or near his team members. He has little or no room for error. That error could result in the injury or death of someone who is counting

on him to produce and who, most certainly, does not deserve to die. He will often have to do this job after waiting for hours or days for the right opportunity or the command to fire. He may have only a fleeting moment to engage his target, and his one well-placed shot could end this whole intense, stressful, and dangerous situation.

> The more accurate the weapon,
> the greater the fear it inspires.
>
> — Lt. Col. Dave Grossman, *On Combat*

"God, I'm Glad I Had a Gun"

In his book *Thank God I Had a Gun*, Chris Bird tells fourteen true stories of effective self-defense with firearms. Some of the stories describe self-defense against intruders in the home, and others describe self-defense against intruders in the workplace. One story that stands out is the description of what happened in New Orleans during Hurricane Katrina. News reports covered the confiscation of arms, the evacuations, and then the roving bands of thugs. Rule of law had ended. Police were ineffective. Those who didn't evacuate became the only "protectors and defenders" still operating. Chris Bird tells of two such protectors who, with the aid of firearms, guarded their own home and family as well as doing what they could to defend the property of their neighbors who had evacuated. The disintegration of society was rapid as law and order dissolved into anarchy. They could not have been successful without their guns.[195]

Man's physical strength limitations led to a need for greater physical force in order to hit an opponent harder and more effectively, resulting in the development of better methods to transfer kinetic energy to an opponent. This process evolved from hitting someone with a handheld rock (providing the momentum energy of more mass than just a fist); to sharp rocks (focusing the energy in a smaller impact point); to a sharp rock on a stick (providing mechanical leverage combined with a cutting edge); to spears (using the latest material technology—flint, bronze, iron, steel—to focus energy in smaller and smaller penetration points); to swords (which permit the option of using a thrusting, spear-like penetration point or the mechanical leverage of a hacking, cutting edge); to the long bow (using stored mechanical energy and a refined penetration point); to firearms (transferring chemical energy to a projectile in order to deliver an extremely powerful dose of kinetic energy).

— Lt. Col. Dave Grossman, *On Combat*

The Angel of the Night
by Lt. Col. Dave Grossman

Fear not the night.
Fear that which walks the night.
And *I* am that which walks the night.
But only evil need fear me...
        and gentle souls sleep safe in their beds...
                because I walk the night.
Carpe Noctem!
(Seize the night!)

Chapter Twelve—Summary

We've discussed marksmanship, field craft, physical fitness, and mental toughness. All of these are skills and attributes learned and developed in hunting. And they can also help prepare us for life-and-death situations. From military and law enforcement to self-defense, hunting has enormous utility to prepare for these circumstances, save your own life, and save the lives of others.

The deep roots of war reach back and are interwoven with our early conflicts with animals. Remember that humanity has been in the *middle* of the food chain for much of our history. Fighting predators and hunting prey would most certainly have shaped every aspect of our conflict with fellow humans. We can only understand our long history of warfare by understanding the hunt and the hunter within each of us. Adrenaline, social bonding, skills, and tools formed our bodies and our cultures.

Within the realm of warriors and hunters, it can truly be said that amateurs talk about "hardware." The vast majority of scholarship and popular culture focuses only on the killing "equipment" in the realms of hunting *and* warfare.

Thus, amateurs talk "hardware," but professionals understand that what is *truly* important, what is *most* deserving of our scholarship, is the "software." The social, cultural, and psychological dynamics that surround the processes of combat and hunting are what really reside at the core of both realms. Thus, coauthor Dave's book *On Combat* has been cited many times in scholarly works, and *On Hunting* is the natural and essential follow-on to the scholarship established in that work.

Warfare can be seen as a sad necessity, but hunting can be recognized as a realm in which we find socially acceptable and beneficial outlets for the drives and talents of combat. The hallmark skills of the warrior *and* the hunter are much the same: marksmanship, fieldcraft, tenacity, resiliency, and mental toughness.

To hunting we can add the exclusive virtues of stewardship, conservation, veneration and protection of nature, and the universal fellowship of *all* who join us in the hunt. These are traits that are worthy of being acknowledged, honored, and sustained as humanity progresses through the centuries and millennia to come.

Chapter Thirteen

The Hardware and the Software

As coauthor Dave is fond of saying, "It's not about the hardware; it's about the software." It's not about using sticks or rocks or spears or bows or guns. It's about the combat mindset and the hunter's heart. It's about the wild brain's intuition and the logical brain's training.

Sublimate, Sublimate, Dance to the Music

Humanity struggles with ancient drives and modern realities. We are designed to feed and breed, to hunt, and to fight to stay alive. Modern life doesn't need quite the same rigor in all of these, yet we still have the drives. So we sublimate.

Sublimation has many definitions, and the one we're using here is this: the diversion of energy from an instinctive drive to a more socially acceptable (even desirable) behavior. It is taking unacceptable urges and drives and making them "sublime." It is "channeling" the energy of our ancient ancestors into our modern life. It is the method used by militaries and police agencies to ensure that each warrior fights but does not go over the line into psychopathy. It is a very fine line.

In police and military work, it has become more and more difficult to keep the line clear. Warriors need just enough "fight" to control the situation, but they must not finish the fight the way our ancient ancestors might have. They need to tap into the ancient brain but keep the Geneva Convention in mind as well.

In personal defense, we know our motivations come from deep within. We know we are training the "fight" mechanism that we are born with. We may take up a martial art…a socially acceptable "sport" that we believe can prepare and protect us. Where allowed by law, we may get a carry permit so we can have a gun with us. We are painstakingly law-abiding about having the required paperwork and training. We probably read books and haunt Facebook pages that tell us how "good guys" with a gun saved the day, prevented mayhem, or protected their families.

> An armed society is a polite society. Manners are good when one may have to back up his acts with his life.
>
> —Robert A. Heinlein, *Beyond the Horizon*

In law enforcement, there is more emphasis on control than on fighting. The deep drives are sublimated into "law and order" and "we serve to protect." When it comes to fight, the use of force continuum describes the guidelines for just how much force may be used in each situation. There is heavy judicial control over most police officers. In one incident, they can be liable for a criminal charge, a civil suit, and an offense under their local police services act. Their drives to hunt and to fight are restrained by the many rules by which they must abide until those drives are needed for a fast car chase and a rough takedown.

Police and military are subject daily to their ancient survival drives. Most of them are proud of this connection. They are proud of being able to take care of a bad situation and proud of being able to show compassion and restraint when needed. Hunters are similarly proud of their ability to take care of themselves. And they're proud of their potential to be upstanding citizens, good Samaritans, and virtuous hunters. This is the "software" that goes with the "hardware" of hunting. And the world is a safer, better place because we do these things.

Let us take a moment, whatever our own personal perspective, and assume the role of the "virtuous hunter." We have sworn the Hunter's Oath. We know that we're driven by our "wild brain," but we strive to find ways to sublimate our drives to more socially acceptable behavior. What does this mean, and what does this look like as it unfolds in our daily lives?

We follow the rules of game management in every detail. We are conscious of the ethics of our hunt and strive to ensure a clean kill in a fair chase. We talk about the wholesome, organic meat with which we feed our families. We are proud of the conservation we support in our own country and in other countries. This is how we convert our hunting drive into a more socially acceptable form.

Hunting in Everyday Life

Those who have not hunted cannot possibly know how it feels. Many nonhunters might wonder if hunting is a natural state of man, but those who've "tasted blood" will tell you that hunting is both man's history and his destiny.

One of the delights of hunting is getting mentally and emotionally lost in nature. It is a total absorption in the present moment. The reality of everyday anxiety fades, and the hunter becomes immersed in the natural world. He feels connected to something much bigger than himself. As Jonathan Haidt put it, we "transcend self-interest and lose ourselves...in something larger than ourselves."[196]

The hunter is in awe of the vastness and power of nature. He feels small by comparison, and he is proud to be a small part of something so magnificent and enduring. This is something the day hiker and the photographer will never experience. The hunter is an active participant in the web of life; the spectator is not.

The soldier feels similarly connected but not quite in the same way. Coauthor Dave quotes a World War II veteran who described it this way: "Once exposed to the heat of [pitched close combat], it welds you into a bonding, not only with friend, but foe, that no one else, no matter how close to you, will ever be able to share."[197]

The hunter also feels connected to his foe, the prey. It is the ultimate bond of sustenance and, therefore, survival. It is something that a nonhunter cannot share. The more magnificent the hunt and the more challenging the prey, the more connected the hunter feels.

Campfire Story: Bill and the Bear

A very good friend of Linda and Keith, and one of our favorite people, is a neighbor of ours. Bill is a decorated and retired police officer who saw his fair share of on-the-job unpleasantries and now lives with his wife in the peace and quiet of a small rural community. Well…it's mostly peace and quiet.

One day Bill was splitting wood in his backyard. While focusing on his job, he was attacked and taken down by a black bear. The bear clawed and bit him on his face and shoulders and tried to drag him away.

It's often said that police officers never retire; they just go off and do something different someplace else. Bill had spent a career being in control of the situation, knowing there is no second-place winner. Winning is the only option. This bear thought he had an easy meal but didn't realize that he had yet to convince Bill.

Bill fought back in the only way he knew. Bloody and torn, he eventually killed the bear with the axe he had been using to split the wood. Bill then drove himself into town where he received medical attention. Bill lived, although with the physical scars of his confrontation, while the bear did not.

Whenever we run one of our police marksmanship courses, we make sure we tell this story every time. Sooner or later, Bill will show up to rub elbows with the blue brotherhood, and we're proud that he does. We always ask Bill to tell the young officers what the moral is that they can all take away from this story. Bill smiles shyly, never one to blow his own horn, and then looks seriously at the class: "Always stay in the fight until one of you is dead. Winning is the only option."

Chapter Thirteen—Summary

In both combat and hunting, it is ultimately *not* about the tools… rocks, spears, and guns. When you dig into the heart of the matter, you find that these endeavors are about mindset and kinship. It is about *bonding* and *thriving* in an arena where innate intuition combines with both life-taking and life-preserving experiences.

Those who have survived combat consistently tell of a similar response. Veterans of the battlefield speak of a "bonding"—of being "forged in the fire of combat"—in a way that few others will ever know or share. The hunter also experiences powerful aspects of that intense bonding.

Human combat grew from hunting, and hunting was refined by combat. It is two sides of the same coin. The oneness with the prey. The intense kinship with your fellow hunters. Even the sensory experiences like auditory exclusion. The more intense the hunt and the more challenging the prey, the more the hunter feels that profound connection with our roots.

And, as we will demonstrate, hunting can be a profoundly healthy part of our ecology and our future.

Section V

It's Who I Am

> Touch the earth, love the earth, honour the earth…
> rest your spirit in her solitary places.
>
> —Henry Beston, *The Outermost House*

If you've never been hunting, you may not be aware that you have a hunting heritage. Even if you have lots of hunting experience, you may not realize how fully and completely you are a hunter. It's in your roots. It's in your brain. It's in your genes.

When we first decided to write this book, we wondered if our research would indicate that we humans have a "hunting gene." Indeed, we did find such indications. And if someone asks us which gene might be the hunting gene, we have to say, "All of them!"

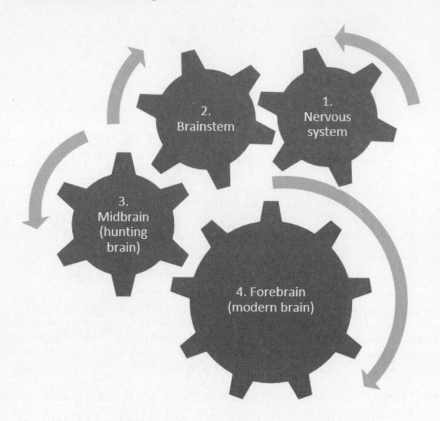

Our brains have distinctly separate components. As shown in the diagram above:

1. Our first "brain" is the fundamental component that most living creatures have in common. It's basically a collection of specialized cells. Its main job is to send signals from one cell to another. It controls things like muscles and glands.

2. Situated upon this primitive networked system is the brain stem. This is a specialized control center that regulates key body functions. It controls our heartbeats, our breathing, our digestion, and our

sleeping. The brain stem also plays an important role in how we operate in the world. It maintains balance and equilibrium, coordinates movement, and manages the transfer of sensory information. In other words, the purpose of the brain stem is to control, to extend, and to protect the very nervous system from which it developed.

3. The midbrain is next, and it is the most complex part of our human brain. Its functions are tightly related to the brain stem functions. It includes the control of the muscles that allow us to speak, laugh, and cry. The midbrain produces the ever-popular dopamine. It also plays an important role in both motor control and cognition. The midbrain is the seat of arousal. This includes fight and flight and mating. It is the center of our motivation and habituation. The midbrain specializes in relaying information for vision and hearing. The midbrain is designed to protect our individual and collective lives: staying alive and breeding. It enhances the functions of the brain stem, projecting the requirements of the brain stem from which it developed. The midbrain is our hunting brain.

4. The forebrain controls most of what we consider "modern human" behavior. It gives us elaborate mental skills that serve our need to stay alive and multiply. It enhances our abilities to think and to perceive. It is essential for helping us to process information and understand language. It lets us develop fine motor skills. The forebrain is designed to make us successful in a "survival of the fittest" competition, which is

"won" by procreating and duplicating our little gene pools. The purpose of the forebrain is to ensure our midbrain operates effectively as *Homo sapiens*. It projects the requirements of the midbrain into our modern world.

When we forget that the forebrain is a servant of our need to stay alive and multiply, we have also forgotten our basic animal biology. If we neglect our basic animal nature, this can lead to a feeling of disjointedness. We have a feeling of not being comfortable in our own skin. We have a feeling of being "out of sync."

> "Natural" man is first "prehistoric" man—the hunter.
>
> —José Ortega y Gasset, *Meditations on Hunting*

Hunters renew their spirits by synchronizing their forebrain with their midbrain, using all their modern human cleverness to exercise (and serve) the midbrain. And the midbrain needs to direct the mechanisms that keep us alive. At the moment of the hunter's kill, the brain reconnects with the brain stem (which interprets and controls significant sensory input).

This may be why hunters sometimes experience phenomena like auditory exclusion. The midbrain is busy hunting and fails to relay the sensory input it has received. As our coauthor Dave says, "I am convinced that auditory exclusion and buck fever are not (just) about the excitement or the stress. I think it is the act of killing and hunting, triggering what I call our 'predator neurons'... That is what makes hunting so intensely satisfying...It is what we were designed to do." This experience during the act of hunting confirms the hunter's humanity.

That, we believe, is what the nonhunter does not see, feel, or understand. While hunters and nonhunters often share a great love of the wilderness and all the wildlife in it, it is this critical difference that separates them. The hunter confirms his humanity in a way that seems exquisitely natural to him, and he sees the nonhunter as a lost soul floating in an artificial life raft.

> It is as impossible to live without reverence as it is without joy.
>
> —Henry Beston, *The Outermost House*

When our forebears settled from hunting into farming, they pushed the wildlife off their land. In modern times, we see this still going on in Africa, where the land is being taken over for agriculture and where the wild plants have become "weeds" and the wild animals have become a nuisance.

As our even more recent forebears settled into cities, they pushed agriculture to the edges of the towns. They covered what was farmland with houses and roads, pushing pig farms and apple orchards outside the city gates.

Like ripples in a pond, for each drop of urbanization, the ripples push agriculture further away and wildlife further still. As cities grow to house our increasing population, the agricultural lands are pushed to less productive land, and our wilderness areas are further compromised. Agriculture (whether pastoral, ranching, grain, or crops) feeds us, and feeds us well, but at a cost. There's always something being displaced and, when pushed far enough, dying.

About 10,000 years ago   About 5,000 years ago   Today

While we may wish that people would value land for its own sake, at the end of the day, the land is only worth what it can produce for humans. It is worth what it can produce as a farm until it is worth the number of high-rises on it.

Urban dwellers tend to see their move to the city as an improvement and consequently devalue the wilderness. Some have no idea why anyone would want to be (or live) in the wild. Even in a country as beautiful as Kenya, the people who live in Nairobi usually don't understand why tourists from around the world come to visit their wilderness.[198]

Everywhere we look, wildland is marginalized, and wild animals are too. If the wild animal isn't worth something, it will be killed. We are all part of it. Even if we live in a city, the land that our house sits on was originally part of the wildland that we took away from the wild animals. The food on our table comes from a farm, and the land that's farmed was part of the wildland that we took away from the wild animals. We push our wildlife back to smaller and smaller parts of the bush, and we leave them the land that's not good enough for farming or convenient enough for cities. And, if they aren't worth enough to our economy or if they are in the way of development, we'll kill them off. We may kill them outright or by habituation (causing them to lose their wildness) or by destroying their habitat.

So, do we just mourn the passing of wildness...in our lands and in our hearts? Are we like the mother gazelle who watches helplessly as her young fawn is attacked and eaten, as the herd moves on without fighting to protect its young? Or will we tap into our predator genes and be the wolf pack fighting for its cubs? As we have demonstrated repeatedly throughout this book, hunting is not the problem. As this vast, global tragedy unfolds before us, hunting can be the solution. Ecologically, economically, psychologically, physiologically, and spiritually: hunting can truly be the solution.

> One of the primary purposes of hunting is to exercise our need to remain a part of the natural world. We still have the desire to participate in the natural process. Our developed world is becoming separated from nature; it is becoming artificial.
>
> Hunting is one of the last ways we have to exercise our passion to belong to the earth, to be part of the natural world, to participate in the ecological drama, and to nurture the ember of wildness within ourselves.
>
> —Jim Posewitz, *Beyond Fair Chase*

Chapter Fourteen

What's Food Got to Do with It?

Much as the modern hunter would like to think of himself as a true hunter, a pure hunter, we must acknowledge that none of us can escape the context of a modern world. We all have modern tools and modern expectations, and much as today's hunter might seek to go hunting to get away from the modern life, he must always return to that world. Everything we do today, hunting included, is put to judgment by our fellow *Homo sapiens*.

The modern hunter needs to be able to defend and promote his pursuit. We've provided many reasons why the modern hunter belongs in the modern world. The world needs the hunter. Wildlife needs the hunter. And the hunter needs to hunt. As Richard Nelson says in *A Hunter's Heart*:

> Most of us are only dimly conscious of our own
> personal ecology—how we're rooted to the soil and
> water, forest and field; how we're affiliated with farm-
> er and fisherman; and how we're connected even to
> wild animals and to hunting.[199]

We hope that we've drawn these connections clearly in this book. But there's also the big disconnect. We need to understand why our nonhunting friends "don't get it." As Nelson adds:

> The supermarket...[creates] an illusion that we can have food without harvest, that life can be maintained without death, that our daily existence is separate from the land, and that we are fundamentally different from all other organisms.

Motivations and Mysteries

In 1980, Yale professor Dr. Stephen Kellert published a well-known study about hunters' attitudes toward nature and wildlife. He divided hunters into three broad groups: utilitarian/meat hunters, dominionistic/sport hunters, and naturalistic/nature hunters.[200] Like most studies of this kind, it's in some ways useful to understand the attitudes, but the naming of the groups tends to divide and polarize. Most of the hunters we know would fall into all three of Kellert's groups.

For the purpose of our study, this book, we have found a much more useful description of why people hunt by talking to other hunters about their hunting experiences. Most are seeking a deep connection to nature, to their food, and to themselves. Many experienced hunters also pass along their knowledge and skill to their children and to new hunters. The experienced hunters want to teach traditions, and they hope the new hunter also learns about the preciousness of life itself. What more could a hunter teach a newbie or a parent teach a child?

Food Defines Us

In ancient times, humans moved toward food. Wherever there was food, humans migrated either seasonally or permanently. The earliest exploration of our little blue planet was likely a reflection of our need to follow the food. Then, with the advent of food "production" (farming), food started to move toward humans. This was the beginning of the end of hunter-gatherers. This was the start of urban centers. This was the start of transportation and trade. Eventually, this was the enabler for huge increases in population. Human relationships with food have defined us for millennia. How we go about getting our food is probably more culturally significant than what the food is.

As Hugh Brody says in *The Other Side of Eden*, "Human beings can reach into themselves and find two versions of life, two ways of speaking and knowing." One is the way of the hunter-gatherer, focused on holistic understanding of his environment. The other is the way of the pastoralist-farmer, focused on controlling and developing the land. "Internally," he writes, "many people are torn between these two ways."[201]

> Every healthy human being has the potential for
> all human qualities; nobody develops one kind of
> strength to the complete exclusion of its opposite. To
> this extent we are all hunters *and* farmers. The differ-
> ences between one kind of society and the other have
> therefore to do with balance.
>
> Without the hunter-gatherers, humanity is
> diminished and cursed; with them, we can achieve a
> more complete version of ourselves.[202]

Many cultures are also concerned about *what* we eat. In the West, we tend to be concerned about a healthy diet. But what we crave is sugar and fat! This is truly normal and a vivid expression of the hunter-gatherer within. From our most ancient days, we burned a lot of energy just getting our food. Finding a rare store of honey or nice starchy tubers or a fat animal was a magnificent treat. Nowadays, we find sweets and fats too easily, and we spend relatively few calories to get them. And meat! The smell of a steak on the barbecue is irresistible to most people, another great expression of the ancient urges within us. We can easily picture the tribe gathering around the ancient firepit to share in the rewards of the hunt.

In the Face of Opposition
In a "deep roots" sense, hunting is not what the hunter does; it is who the hunter is.

> "Why do you hunt and fish?" I'm often asked. The easiest answer is: "My father and all my ancestors did it before me. It's been part of my life since childhood, and part of my identity, like being a southerner or a Baptist."
>
> —Former U.S. President Jimmy Carter, "A Childhood Outdoors" in *A Hunter's Heart*

In *A Hunter's Heart*, Thomas McIntyre notes that, in the face of adversity, hunters persist in the fight for the right to hunt. He asks, "What is it in hunting that can create such defiance and determination? What can it be that this hunter…knows?"[203] It's one thing to look at pretty pictures or cute videos, and it's another to have a deep understanding of the total worth of an animal on this earth. If you think the wild animal's value is simply to be cute and entertain your heart, you've missed what the hunter sees.

The hunter is, at his core, in his heart and soul, a conservationist. And the conservationist knows that all wild animals die. He knows that the environment has a "carrying capacity." And he knows that most of the wildland left is the poorest land on earth (not suitable for either cities or farms). The hunter-conservationist knows that protein is what fed our quickly developing brains. It was protein that enabled the parts of the brain responsible for language, logic, and art. Our connection to animals is in our hearts, our minds, and our very muscle tissues. We believe it is disrespectful to not let prey be prey. Taking away their "prey-ness" is taking away their identity, their way of living, and their purpose in the higher good.

Thousands of years ago, when agriculture started to make its mark on the landscape and in the human heart, we easily traded the way of life of the hunter for the way of life of the farmer. It was not all sweetness and light. The need to cultivate and develop property led to boundaries and to fighting to maintain boundaries. The nature of farm work required all the family (and large families were an asset) to toil for long hours. The families were dependent on the whims of the weather and could suffer greatly, for example, in the case of drought and concurrent famine. However, agriculture's relatively enormous production of food (and the ability to transport it) fueled the increase in population and the increase in urban centers. Indeed, these (food production and transportation) are probably the main cause of the population increases that took us from a few million to just over a billion by 1850. Improving technology and medicine likely were the significant drivers that took us to nearly eight billion today.[204]

Still, urban dwellers often have a strong yearning for things rural. The higher-priced homes are those near open land, and urbanites flock to farm-experience events like sugaring off in the spring and apple picking in the fall. Big-city zoos and small-town animal farms abound for tourists to visit, and everywhere in the media (news reporting, movies, and advertising) we see the glorification of country living. This, however, is the "Disney" view of life in the country and miles away from reality of life in the wild.

As Paul Shepard points out, we modern urban folk have bought into a myth about rural life and peasant values. While our hearts yearn, our minds need to be better informed. Shepard adds that when we are picturing a "return to the farm," what our hearts are truly yearning for is a return to solitude, clean air, and clear water, which are not made by agriculture but are the properties of the natural earth.[205]

Why do we need to be better informed? Because the wild animals' lives depend on it. The ecological and economic salvation that hunting can contribute, Namibia versus Kenya, has been mentioned before, but it is essential and appropriate to constantly fold this critical concept back into the narrative. Hunting is the key if we want to have animals in some place other than zoos and petting farms: the *wild* animals' lives depend on it.

The Hunting Culture

There are some who can live without wild things,
and some who cannot.

—Aldo Leopold, *A Sand County Almanac*

Modern hunters identify with the hunter-gatherers of the ancient past.

Thus, the conflict today between hunters and antihunters can be understood as a conflict of wild heritage (and birthright) versus modern advancement (and privilege). The hunters started the conservation movement over a century ago, trying to ensure there would be a place for wild game to live and thrive. Many of the recently "woke" conservationists have forgotten (or never knew) the roots of the movement. And while they have space in their hearts for habituated game (photo safaris, game parks, and the like), they have little space in their hearts for wild game or the wildness within us.

The urbanized and cultivated man has almost always felt a
funny snobbery toward anything wild, man or animal.

—José Ortega y Gasset, *Meditations on Hunting*

Zero Degrees of Separation

> Rather than creating personalities and world views,
> hunting merely reflects them.
>
> —David Peterson, *Heartsblood*

While doing the research for this book, we were surprised to see that many of the cultural characteristics of the modern hunter are similar to the culture of the ancient hunter-gatherer (at least as far as anthropologists can divine). One of the most striking things we noticed was the way people think about the word *primitive*. The general population often defines *primitive* as "lacking in sophistication, crude, rough, or uncivilized," whereas modern hunters tend to see *primitive* as meaning "primary, basic, or fundamental." There are zero degrees of separation between the hunter and the reality of the wild. Kay Koppedrayer confirms this in *Hunting— Philosophy for Everyone*:

> Instead of a denigrating term referring to a way of
> life or a time that people grew out of as civilization
> advanced, "primitive" as used in bow-making circles
> implied a hyperreal state, one that enables the use of
> all senses without the intrusion of other mediating
> forces…"Primitive" thus becomes a means of cele-
> brating an unencumbered life.[206]

Hunters have great respect for *Homo sapiens'* deep roots, history, and mythologies. The hunter cherishes and savors the idea that his lineage goes back to ancient times. The hunter has accepted the role and responsibility to be the guardian of humanity's hunting heritage. And it is through the hunter that we honor the origins of humanity. Indeed, Mary Zeiss Stange writes about this primitive link being part of "hunter consciousness" that spans time and circumstance and unites hunters through history and around the world. She writes:

> [Ted Kerasote] is not talking about so-called hunting with a camera…or about "ship-in-a-bottle" environmentalism; he is talking about the forest his very life depends upon. This is not to say that only a person who actually hunts for food can listen in this way; but it is to suggest that this mode of hearing arises from—and is in some ways primordially linked to—hunter-consciousness.[207]

Conservation Is Our Nature
In the *Washington Post*, journalist Steve Sanetti writes:

> Today's green movement uses certain buzzwords—organic, locavore, renewable—to the wry amusement of 15 million to 20 million of us who've actually lived the eco-friendly lifestyle that these words describe.[208]

We are hunters. Hunters take great pride in their skill and traditions. They like knowing that only one or two people have ever touched their meat, from field to table. They enjoy meat that is fully and completely organic as well as being lean and healthy. They know that they help the economies of otherwise poor areas. They support game laws because they know this helps the game species (as well as nongame species that share the same habitat).

One with the Land and the Clan
Culturally, hunters are tied to the land. They love the earth, the waters, and the creatures who dwell with us on this fragile planet. They are knowledgeable about nature and the wildlife that share it with us. They learn about the game animals they hunt and the

creatures' habits and needs. In addition to advancing their skills with the technology they use, they also develop techniques and tactics that are independent of any technology. Hunters work on developing physical strength and endurance so they can hunt effectively and ethically. They understand that they are alone with their ethics when there are no witnesses to their behavior. We demonstrate our "character" by what we do when no one is watching. Hunters are proud to live up to this test of virtue. Hunters are proud of their mental skills. They are pleased about their ability to stay alert through hours of patient observation. They are especially happy that they can control their minds and marshal their thoughts when the moment of heart-pounding thrill arrives.

Hunters may hunt alone or in a small band. Social anthropologists estimate that ancient humans banded together in groups of about twenty-five individuals. Some Indigenous North American tribes say that the hunter must share his first kill as broadly as possible. Most hunters we know share within the family and the community, especially with nearby hunters who were not lucky in the hunt and, where allowed, with the local food bank.

As Jonathan Haidt puts it in *The Righteous Mind*:

> We live most of our lives in the ordinary (profane) world, but we achieve our greatest joys in those brief moments of transit to the sacred world, in which we become "simply a part of the whole."[209]

The hunter's sacred world is "nature" or "the land" or "the wilderness." He achieves his greatest joys in those brief moments when he becomes one with the wilderness. Hunters like other hunters and bond easily with these kindred spirits. They all value self-reliance, resourcefulness, and initiative. They all have their own simple, heartfelt rituals. The woodlands are a cathedral, and they honor every animal that nature brings to them. And they all are grateful—profoundly grateful—for the grace of meat on their table. But more importantly, most of all, they are grateful for the experience of their wild lives.

The New Elite?

There is a world of difference between people who believe that "zero degrees of separation" is a good thing, a natural thing, a return to ancient roots and people who measure their success by material things that move them further from the source. For example, hunting to put meat on the table is regarded as "primitive," whereas eating well-aged beefsteak prepared by other hands is seen as "fine dining." Urban life is now so far removed from wild things that, for most people, a trip to the zoo is seen as a visit to the wild. The authors of this book are all hunters, and they have found a powerful common experience, an experience that practically every hunter has shared. When we took our grandchildren to the zoo, those little ones were excited to see the animals. But we had to turn our eyes away in sorrow tinged with disgust. How we wished we could take those precious grandchildren to the Kalahari to see giraffes and kudus in the wild.

Like all hunters, we yearn for the day when we can take these little ones hunting. We want them to see the splendor of black bears and white-tailed deer in the true wilderness. We want to plant their tender young roots into the rich soil of our primordial past. And we want to teach them to become a truly connected part of the wild. In the end, this may be the hunter's greatest joy, the ultimate fruit of a lifetime spent striving to know and preserve this way of life: to pass the joy and wonder and majesty of it all to a new generation.

We do not want to take them on a photo safari, where largely habituated animals are visited by eco-tourists in gas-guzzling buses, whose schedules are synchronized with feeding times and who are mostly interested in ticking the boxes on their list of animals seen. This is perhaps one step better than a zoo but nowhere nearer to reality. There are other forms of nonhunting safaris that provide closer contact with wilderness and wildlife (which are usually called adventure safaris or similar). Hunting is usually the closest reality modern humans can hope to see even though we are but visitors to the wildlands.

There are very few wild places and very few wild people on this earth, but those that remain are viewed as curiosities. They are small tribes in places like Borneo or deep Africa, and when there's an encounter with them, the general feeling of the public is "Wow. How did they get left so far behind?" But the hunter knows the real question should be, "How did *we* get so far away from our origins?"

Damn That Rear-View Mirror

While hunters may identify with their ancient hunter-gatherer ancestors, for most modern people, it's all an inconvenient truth. They would rather deny their roots than acknowledge that, at one level, they are another animal, albeit one with a pretty fancy brain. They would particularly like to ignore the fact that the fancy part is not in complete control and that the oldest part of the brain still controls the show.

Sometimes, we feel an urge we don't really understand. It comes from that old way of life, that primitive part of the brain. The modern brain doesn't understand because it's truly not part of that world. For example, the hunter's heart yearns for solitude. Most hunters don't want to take their animal on the first day of the hunt because it ends the process, and they want to keep hunting as long as they can. We believe these feelings come from our old brain.

Chapter Fourteen—Summary

We can never completely escape from our twenty-first-century world. Even if we immerse ourselves in the most ancient hunting technology and techniques, we must inevitably operate within the realm of modern humanity with its technology and expectations. Thus, we must always be prepared to justify the act and art of hunting to our fellow *Homo sapiens*.

In order to preserve nature's ancient gift of hunting in the modern world, we must clearly explain and justify who we are and what we do. It is essential to understand that the world *needs* hunting and the hunter.

The preservation and future of wildlife and wilderness depends largely upon the hunter. Only the hunters in their vast numbers—who willingly pay for the privilege of becoming a direct participant in the food chain—can and will sustain the wilderness in the modern ecology.

And the hunt is essential to the wholeness and wellness of the hunter. It is not just what we do but also who we are. Hunting is a taproot that connects humankind to ancient gifts of independence, ingenuity, and initiative. Goethe said that "Nature is the visible garment of God," and the hunter becomes a thread in that blessed garment. Wilderness is the hunter's cathedral, and he venerates every living creature in it. All humans eat, but only the hunter is truly in tune with his nourishment. We revere those who give us their lives so that we may be nourished. Perhaps most of all, the hunter is deeply grateful for the wilderness and the gift of each creature that shares its life with our life.

To those who do not know these gifts, we say, "Come. Join the hunt." Touch your ancient roots. Nature needs you, and you need nature.

Chapter Fifteen

Beware the Righteous Man

Hunting is in our genes. Just as a domestic cat will pounce on a laser light, we are irresistibly drawn to *la chasse*. Just as a hawk will alert to the slightest movement of its prey, we are naturally attentive to an unexpected movement in the field. Just as a hound or retriever puts its nose to the ground, we judge our surroundings by the smell in the air.

We have all the impulses of our ancestors. Like prey, the hair rises on the back of our necks when we feel threatened. Like scavengers, we focus intently on a treasure hunt, and we love that "something for nothing" feeling. Like predators, we love to follow a path that leads over the next hill to see the next valley.

Hunting can include all these feelings, but most significant of all is the thrill of the kill. Hours (and days) of patience, of enduring cold, of staying silent, all for the momentary rush of capturing our quarry, followed by a mixture of sadness, happiness, and satisfaction. We believe what's happening is that we are resynchronizing our ancient brain with our modern brain. The unity we feel at that moment is the unity of our brain. We are "in sync" with our ancestors' heartbeats. We are resetting the relationship between the forebrain and its boss, the midbrain.

The forebrain's job is to do everything it can in the modern world to protect and develop the midbrain. The midbrain's job is survival: fight or flight, feeding, and breeding. And never doubt that the midbrain is the boss. The forebrain is that "thin veneer of civilization." If you think this doesn't apply to you, think of the last time someone cut you off in traffic. That little flash of road rage gave you a glimpse of how thin your "veneer" might be. For an instant, you saw the "fight" mechanism of your midbrain. If you've ever been in the bush and realized that a bear is coming up your back trail—that is, the bear is tracking you—you might have a glimpse at the "flight" mechanism of your midbrain. Or maybe you're a city-dweller and you heard footsteps following you in a dark parking garage. That quick, strangled feeling you get in your chest? That's your "flight" mechanism too.

So, how does this connect to hunting? Well, there's a hormonal connection, the hunter's feeling of unity at the moment of the kill. It's in his hormones. It's in his biochemistry. It's in his genes. As more human genetic research is conducted, we are finding out how closely related we are to our ancestors. From our most ancient days, all of us were hunters. It is reasonable to expect that our genetic configuration would enable us to thrive as hunters.

It is common for all people, including those born and raised entirely in cities, to dream often about animals. Children learn about both biology and sociology through animal stories. Small children imitate animals. Many children's games are reenactments of predator-prey situations. Hide-and-seek is an obvious hunting scenario.

Recently we saw a nature show on TV where a cheetah mother was teaching her three youngsters to hunt. She put an impala lamb in front of them. When the lamb froze in place and didn't run, the cheetah youngsters didn't know what to do. When the lamb ran, they immediately started the chase. It's the hunting instinct…and we have it too.

So, we must ask, like the tamed elk who loses its "elkness," are we tamed humans losing our humanity?

That's All Well and Good, but What about Nonhunters?

So, can nonhunters resynchronize their brains?

Well, they can try.

Sometimes, nonhunters fear hunters. They say it's because the hunter can kill. The hunter has the tools. They have the skills. They have the mindset. So, what's to stop hunters from becoming murderers? Hunters try to empathize with people who feel this way, but they end up finding them tragically uninformed and out of touch with the reality that hunters cherish.

The truth is just the opposite: statistically speaking, the hunter is far less likely to commit a violent crime. Some major hunting states in the US like to proclaim that during the hunting season, with all those armed hunters in the woods, they represent the second (or third or fourth) largest army on the planet!

> You cannot invade mainland United States.
> There would be a rifle behind each blade of grass.
>
> —Admiral Yamamoto

In the same way, fatalities from hunting accidents have become extraordinarily rare. In Canada, probably the major cause of serious injuries and deaths during the hunt is now heart attacks. It used to be falls from tree stands, but education (and good harnesses) has changed all that. It is worth taking a minute to explain the revolution in firearms safety that has made hunting accidents so rare. This revolution is best explained in the "Four Universal Gun Safety Laws."

- First, treat every gun as if it is loaded.
- Second, point the muzzle in a safe direction.
- Third, keep your finger off the trigger until your sights are on the target and you've made the decision to shoot.
- Fourth, be sure of your target and what is around and beyond it.

There is no need for nonhunters to fear hunters, certainly less so than they might fear visiting Chicago or driving on our highways anywhere. At the end of our analysis, we find that the root of their fear is wrapped up in their denial. It's not that we can kill. It's that we hold a mirror to *their* souls. They are afraid of the dark spaces in their own hearts. And as such, we believe, they are afraid of their own humanity.

Embracing Nature

Hunters and many modern nonhunters embrace nature. Many people simply visit the country to recharge their energies. They visit a farm, pet the tamer animals, maybe feed the goats, or ride a pony. They buy some organic maple syrup and some homemade cinnamon buns. And then they go back to the city, perhaps thinking, *Oh, I'd like to live like that all the time!*

The farm may be organic, but it is not wild. Even the "organic" movement may have less to do with nutrition and more to do with an unconscious need to return to roots. But the real roots are deep and covered in dirt.

Living Vicariously

Many modern people experience nature mostly through their imaginations: while reading, looking at pictures, and watching videos. Indeed, we experience a large part of our modern lives through media. And often that experience is more influential in shaping our behaviors and beliefs than is living hands-on with reality.

For example, there is a large variety of hunting and outdoor television networks and many hunting shows on those networks. Hunters are the major audience for these TV shows. Often the shows depict safari hunting in Africa. Hunters love to imagine that they are there, experiencing the hunt! Their reality is grounded in their own safaris, and their experience is broadened by watching the hunts of others. Having real hunts in his background, the hunter can distinguish the erroneous depictions of reality and still vicariously enjoy the experience of other hunters.

Nonhunters often experience nature through media too. But they often do so with no grounding in reality. They may believe that the hunter can walk into the woods and shoot anything he wants, all within the half-hour show time. They may buy into the Disney version of nature, where the prey would never die if it were not for the mean old hunter. They may think that, in fact, old animals die peacefully with their families gathered around. In reality, there is no palliative care in the wild. Animals either starve to death because their teeth are worn down and they can't eat, or they are ripped to pieces, usually while they are still alive, because they can't defend themselves from a pack of wolves or a pride of lions.

And when the wolves and lions die? What if there is no predator to give them a comparatively quick death? Then they die the slowest and most hideous of all deaths, from dehydration or by being eaten alive by insects and rodents over a timeframe of days.

Every creature in nature dies. Those who die of old age die the worst deaths. Those who die from the hunter's bullet, at the end of their lifespan, have received one of the greatest gifts that humankind can bestow upon nature.

And while there's nothing wrong with living vicariously from time to time, it is wrong to think it replaces reality.

Modern people are driven by a continuum of effort that requires them to domesticate (dominionize) every creature on the planet...to include wild game, and the wildness in the people who hunt wild game.

—Mary Zeiss Stange, *Woman the Hunter*

Protesting Reality

There is no doubt that our current view of the world is shaped by our contact with it. George Sessions, American ecologist and philosopher, (in the foreword to *The Tender Carnivore and the Sacred Game*) emphasizes in agreement with Shepard that the tragedy of modern civilization is the failure of our youth to "ritually bond with wild nature."[210] He writes that others have noted that the natural world plays a crucial role in the psychological development of children. Sessions quotes historian Morris Berman, who asks, "How...could psychologists have missed this?"[211]

Indeed, we believe it is all too easy for psychologists to have missed this because they did not have nature as a part of their own reality. They have a blind spot, and they can't see their own blind spot. Now with a multi-generational blind spot, we perpetuate the problem without seeing it.

Paul Shepard describes the pre-agricultural human as having "achieved a rich humanity long before metals, pots, wheels, kings, and theocracies appeared."[212] Shepard goes on to say that the history of civilization can be seen as an endless, tragic, senseless "persecution of non-farmers...a war on hunter-gatherers."[213] Shepard refers to a cultural hatred of hunters and concludes that "genocide began on earth with the extermination of hunter-gatherers."[214] Importantly, the father of the modern conservation movement asks, "Can we face the possibility that hunters were more fully human than their descendants?"[215]

Controlling Our World

Hunters have the power of death. They completely understand that while, in any one situation, it's "eat or be eaten," in the long run, it's "eat *and* be eaten." As David Petersen writes, "Psychologists would say that denial of such a fundamental guilt—that another life has to end in order for one's own to continue—sets the stage for self-righteousness."[216]

Pathologies of Abstaining

One of the great pathologies of modern times is an abandonment of our predator roots. Generally speaking, we've allowed the forebrain to dominate our lives. We deny the existence of—or at least the significance of—our midbrain in our day-to-day lives. We hear the midbrain cry out, and we channel its impulses into a more socially acceptable form. We intellectualize every aspect of our lives, rationalizing as we go. Fight-or-flight has become an entertainment found in sports and video games. Feeding is a social event, not a survival event. And breeding…well, it's mostly a commodity, over-marketed to a tired population looking for thrills.

Most people have many degrees of separation between the fundamentals of life and the life they lead. This skews their view of how the world works and gives them a chronic form of "separation anxiety." They don't know what they're separated from, but they know they feel unattached from something. We believe this is the pathology of abstaining, abstaining from the very grounding activities that re-synchronize the parts of the brain…like hunting.

Righteous Denial

Denial is one thing, but denial threatened by reality often becomes righteous denial. There's a lot of that going on these days. People who are simply in denial usually see others' views as harmlessly wrong and can practice some variation of "live and let live." Although they may disagree with an idea or a practice, they can carry on with their own way of life without further ado. People who are in righteous denial feel a need to correct others who have a different point of view and may feel obliged to harass, bully, or even annihilate them.

> The hunter, said the ancients, is not the one who will be found wanting in due reverence to the gods. The hours spent angling are not reckoned against a man's life span. "For," says Xenophon, "all men who have loved hunting have been good." The voices that have been raised against us have been charged with emotion, coming, not as they pretend, from philosophical heights, but from tight compartments, insulated from reality. For this reason, sportsmen have scorned them.
>
> —C. H. D. Clarke, *Autumn Thoughts of a Hunter*

Land Connection

For most of us, our connection to the land is paved with asphalt. If we visit the "wilderness," it's usually a park of some sort. We drive there. We bring a backpack with a water bottle from home. We hike through the "wilderness" on marked trails, looking at the scenery. We enjoy it very much and feel a momentary break from the daily grind of our urban lives. This is a connection, though not a deep connection, to wildness.

> One day at a summer camp, where the women were baking bread in ovens they had scooped out of sand heated with large fires of driftwood, Mary Adele said: "On the land we are ourselves. In the settlement we are lost. That was the way they made our minds weak."
>
> —Hugh Brody, *The Other Side of Eden*

Nature is a background to our human activities. It is our play area. It's expendable. We can fence it in, pave it over, or put houses on it whenever we wish. If we view the wilderness through our modern forebrain, our starving midbrain gets no nutrition from the expedition. As Richard K. Nelson says in *A Hunter's Heart*, our forebrain believes the illusion "that we can have food without harvest, that life can be maintained without death…and that we are fundamentally different from all other organisms."[217] But our midbrain is not fooled.

Few are the people familiar with the age-old bonds to the land. So many are those who see nature through the mind-numbing fog of a picture tube. Nature is to them a game show. It has no reality. It is something that exists in the fantasy of television. There is no blood and gore and life and death to it. Television sterilized the savagery of nature, and the logical reaction of people exposed to this subtle propaganda is a growing desire to sterilize—neuter might be an even better word—the pagan hunters and fishermen.

—Craig Medred, "Venison Sandwiches" in *A Hunter's Heart*

Many people think that visiting nature, and especially visiting nature with a camera, provides a connectedness to wildness. Not only is that not true, but *the illusion that it might be true* is dangerously wrong.

In an essay called "The Camera or the Gun," included in *Hunting—Philosophy for Everyone*, Jonathan Parker emphasizes that, while a photo safari can be an enjoyable thing to do, it in no way connects the participant to the environment. The photographer is truly a spectator rather than a participant. Mistaking this is dangerous because, as Parker says, it can make us want "to remake the natural world." He says we need the insights of hunters who alone have "this profound and deep access to environmental truths and mysteries," and photographic hunters cannot replace them.[218]

[Ortega] condemns people like me who have substituted looking for hunting. We are, he says, voyeurs—still hunters, but only idealistically, platonically; we are guilty—the ultimate sin in his view—of affected piety. I have thought about that argument a lot. I want to refute it, but I confess that I am unable to say where the error lies.

—Paul Gruchow, as quoted in *Heartsblood*

The connection the hunter seeks is not simply a visit to the bush. The hunter seeks to reaffirm his own wild humanity by vividly, strongly uniting with wildness and wild animals. It is their own individual solution to the soul-less anguish of "civilized" life. They aren't working to solve everyone else's problems, only to dissolve their own feelings of separation, separation from nature. When stalking an animal, the feeling of separateness melts, and the hunter is bonded to all of nature, a unity that includes his environment, his prey, and his whole self.

When we hunt, we enjoy every moment, and 99.9 percent of the moments do not involve a kill. When we walk in the bush, going to the stand or while stalking, we become immersed in the environment. For a time, we have no gender, no age, no conscious thoughts. We have only an awareness, both broad and specific at the same time. Our pack weighs nothing, and our bow rides easy on our back. We are enmeshed in nature, truly an integral part. It's an awesome feeling. Even without capturing our prey, we return often to the woods, just for a chance to feel that feeling again.

Mortality...and Edibility

It is a common observation in psychology that many people fear in others what they most fear in themselves. Many times, in discussions with nonhunters, it soon becomes apparent that they are uncomfortable about the whole idea that life requires us to kill the things we eat. While they don't usually see any problem with killing vegetables and grains, they don't want to think about how many animals die to protect those vegetables and grains before they get to the store. Many people are uncomfortable about killing animals for food. While they don't mind eating meat, they certainly don't want to be involved in the killing process. When asked, many would have to admit that, at the moment of truth, they don't think they'd be able to pull the trigger or release the arrow. In other words, without someone else to do their killing for them, they know that they cannot survive.

Even the vegetarian and the vegan survive only because we exterminate millions of rodents around our granaries, thousands of rabbits around our vegetable crops, and hundreds of wild pigs around our peanut farms every year. This is the fundamental truth from which they flee. At some level they know that to live is to kill, that their life apart from killing is, at best, a fantasy (and at worst, a lie). And they know that they cannot survive without other people to do their killing for them. And thus, they fear in others what they most fear in themselves: the truth they deny is the need to kill in order live.

Hunters, of course, fly in the face of this denial. They not only willingly cause the death of an animal, but they also celebrate the food it provides. The hunter has made peace with death. He sees it as an inevitable consequence of life. Life's greatest challenge, the final frontier, the ultimate threat, is death. And the hunter has made peace with death and thus a true peace with life that the vegetarian will never know.

The hunter has taken the role of predator and lives it to the fullest. And that's probably the final straw for both nonhunters and antihunters alike. Reminding them that we humans are predators also reminds them that we have been (and could yet be) prey. In the essay "Flesh, Death, and Tofu," T. J. Kover sums it up this way:

The frequent references by vegetarians to meat eaters as "flesh eaters" or consumers of "rotting corpses" reveals a fundamental disquiet among vegetarians concerning the basic connection between life, flesh, and mortality. Yet perhaps the most obvious example of this connection is the frequent rhetorical question asked by vegetarians: "How would you like to be killed and eaten?" This capitalizes on one of the central reasons behind our unease in being connected to the natural world—*our edibility.*[219]

Circle of Life

We Westerners often think of life versus death, or predator versus prey, or mind versus body. This dualistic way of looking at the world sometimes makes it easier to understand, but like any model, *it's not reality.* Reality is messy.

In his essay "From Arctic Dreams," Barry Lopez describes his attempt to understand one of the few remaining Indigenous hunter cultures. He tries very hard to understand (and to be sympathetic), but he clearly is in "culture shock," and he fails to see the whole truth of the hunter culture. He writes, "No culture has yet solved the dilemma each has faced with the growth of a conscious mind: how to live a moral and compassionate existence when one is fully aware of the blood, the horror inherent in all life, when one finds darkness not only in one's own culture but within oneself."[220] This is the full irony of the hunting situation. Lopez is a spectator. He sees compassion and killing as a paradox. It is only a paradox to a modern Western dualistic culture. To the rest of us, it is part of the whole, the continuum, a seamless sustaining reality.

In contrast, Brian Seitz, in an essay called "Hunting for Meaning," writes that he is aware of "how much life follows a hunting death, the bodies of game animals becoming human bodies, deer literally transubstantiated into women and men."[221] This resembles the hunter-gatherer belief that we become what we eat.

> I remember my young niece (about six years old at the time) asking me if it was really true that I hunted deer. "How could you kill them?" she wanted to know. "How could you eat them?" I replied that many cultures believe that we become what we eat. If we eat the bear, we become strong. If we eat the deer, we become swift of foot.
> She thought about that for a moment and smiled, "And look what we eat…we eat pigs and chickens!"
>
> –Linda K. Miller

Where the Long Shadows Grow
by Linda K. Miller

The sun hangs low
And the long shadows grow.
The hunt is done
And with the setting sun,
A soul moves on
And a life is gone.
My heart is torn…
From this sadness born
A flush of bliss
Like summer's kiss.
I thank my god
With a prayerful nod;
I know some day
I'll be called that way.
It'll be a while
Before I'll walk that mile
Till my sun's low
And the long shadows grow.

Are We beyond Redemption?

While we watch the misguided attempts to squelch our ancient drives, we mourn the death of the spirit within. But are we beyond redemption? Or is there still hope that we can reunite who we truly are with how we act in the modern world?

> Despite our ever-changing, ever-indignant world with its growing ignorance of and indifference to the ways of the wild, I remain a predator, pitying those who revel in artificiality and synthetic success while regarding me and my kind as relics of a time and place no longer valued or understood.
>
> —M. R. James, "Dealing with Death" in *A Hunter's Heart*

What options do we have to restore our wholeness, to reunite our twenty-first-century brain with our inner hunter? We can deny our inner hunter, or we can sublimate our natural urges to more socially acceptable habits. Both these alternatives are insufficient in the long run: both leave us conflicted, unsatisfied, and at risk of dulling the very things that make us human.

There have been changes in the hunting landscape in the past one hundred years. Both hunters and nonhunters are more aware of the need to protect not only the wild game but also its habitat. Both hunters and nonhunters work (in their own ways and usually separately) to protect and improve ecosystems, habitat, and corridors.

But there's a destructive force at work too. We are watching a tragedy unfold: it is the success of the "animal rights" movement. As David Petersen cautions us:

> Were animal rights advocates ever to realize their
> fantasies of a world without predation, elk and deer
> and all the others would quickly lose the razor's-edge
> wildness that informs, enables, and defines them.

Petersen goes on to say that half-tamed or "habituated" animals are at risk of physical and intellectual decline. He emphasizes that this is on its way to being a disaster for the animals and for humans:

> Clearly, elk are better off hunted than housebroke,
> which kills the elkness in them, reducing magical
> wild things to neighborhood pests.[222]

Hunters can see where our fellow man is going astray. They feel it, understand it, grasp it in their own hunters' hearts. They continue to fight an honorable campaign of promoting awareness and understanding. Although some days the task might feel overwhelming or insurmountable, the hunter must maintain strength, hope, and the warrior attitude. In the end, we must embrace the hunter in each of us and protect the wild, both in our ecosystems and in ourselves.

The Special Effects of Media

> Those who can make you believe absurdities
> can make you commit atrocities.
>
> —Voltaire

The media has built its own "reality," a virtual reality, that doesn't reflect reality. Rather, it reflects the thoughts and feelings of those who filter reality to our books and to our screens. The media float over reality and add strength to the currently popular narrative. Communications guru Marshall McLuhan once said, "The medium is the message." If you learn about nature through a medium such as television, film, the internet, or video games, the message you are getting is about the medium you're using, not about the subject (nature). Once the full truth of McLuhan's message is understood, you know in your heart that to learn about nature, you must go into nature with nothing mediating your connection to nature's reality.

Through the media, there is a "widespread sentimentality that projects meanness into carnivores and kindness into herbivores."[223] In fact, in order to maintain a constant population, each animal needs only replace itself. All those other babies can (and will) die unless that habitat can carry a larger population. With turtles, that's scores of babies each and every year that die in the wild and could be hundreds or thousands over a single mother's lifespan. The rest of the babies feed other wild animals. As Shepard says:

> That animals live by eating other creatures, that more
> than 50 percent of the newborn will not live a year,
> that predators are good and necessary are incompat-
> ible with the sentimental cloak that characterizes the
> nineteenth- and twentieth-century animal story, the
> Walt Disney view…That Disney films often show the
> predator as necessary to biological systems does not
> change their essential conceit—that at best predators
> are necessary evils. Sympathy is always with the prey,
> which usually escapes in the depicted episodes.[224]

"Sympathy is always with the prey." After decades of this indoctrination, it is no wonder that we now have a generation of young people who strongly identify as victims.

Shepard says that by hiding these realities from children, in hiding the killing of some animals by others, we hinder children's brain development. Without the outer knowledge of reality, the child's inner monsters become bigger than life, and his only response is to repress them. As Paul Shepherd says in *Thinking Animals*, "The child's first and most powerful attention to the surroundings, apart from his human family, is riveted upon animals."[225] He spends his first years understanding the taxonomy, order, and structure of the world by first observing, naming, and sorting animals. He learns about himself by identifying what he has in common with animals and, importantly, identifying how he is different.

> Humans are biologically primed to seek survival data,
> and violence is the ultimate survival data.
>
> — Lt. Col. Dave Grossman, *On Combat*

In *The Science of Fear*, Dan Gardner writes about the ancient structure of our modern brain and how our perceptions can be twisted by the media. He's quick to point out that he's not "media bashing" but that the members of the media have the same ancient brains. Here's how he says it works:

> Our ancestors didn't jump up and scan the horizon when someone said there were no lions in the vicinity, but a shout of "Lion!" got everyone's attention. It's the way we are wired, reporter and reader alike.[226]

The Ambivalence toward the Original Hunters

> To date, the hunting way of life has been the most successful and persistent adaptation man has ever achieved. But what's left of the original hunters are mostly tribes pushed to the fringes of their original range…or integrated (usually poorly) into the agricultural or industrial edges of the mainstream culture.
>
> —Lee and DeVore, *Man the Hunter*

We could learn a thing or two from the hunter-gatherer tribes that still exist on this planet. We could learn about a deep understanding of nature and ecology. We could appreciate music and dance that connects generations. And we could better understand the tradition of marking the transition from youth to maturity with a period of time alone in the wild. If we choose to follow the path of the hunter, then we can follow their lead in respecting animals. We can sharpen our wits by engaging the noble game animal. We can show respect and thanksgiving for its death. And we can learn to use all parts of the animal. Let us picture the proud people that these hunter-gatherer tribes were before contact and still strive to be.

> We adjust our histories in order to elevate ourselves in the creation that surrounds us; we cut ourselves off from our hunting ancestors, who make us uncomfortable. They seem too closely aligned with insolent, violent predatory animals. The hunting cultures are too barbaric for us. In condemning them, we see it as "inevitable" that their ways are being eclipsed.
>
> —Barry Lopez, "From Arctic Dreams" in *A Hunter's Heart*

Hunting Trophies as Symbols

> The true trophy hunter is a self-disciplined perfectionist seeking a single animal, the ancient patriarch well past his prime that is often an outcast from his own kind…If successful, he will enshrine the trophy in a place of honor. This is a more noble and fitting end than dying on some lost and lonely ledge where the scavengers will pick his bones, and his magnificent horns will weather away and be lost forever.
>
> —Elgin Gates, *Trophy Hunter in Asia*

We read an interview with a hunter who was trying to understand why some people are so against so-called trophy hunting. "What's the difference?" he asked. "I hunt for trophies and eat the meat, or I hunt for meat and keep the trophies." Indeed, what *is* the difference?

Trophy hunting is a term best left for antihunters. All hunting includes trophies. They are the nonedible parts of the kill, the hides and horns, the bones and teeth. A long time ago, the animal pelts may have been used to provide camouflage for the next hunt or provided teaching props used by older hunters. Hides and furs provided the clothing and bedding that allowed early humans to live in climates that were colder, nearer to the edge of the retreating ice of the Ice Age. Bones and horns were used for tools. Tribes used pelts and heads for ceremonies, while fangs and claws became jewelry and totems. For the hunter, a trophy is an icon of remembrance. As Paul Shepard points out, taking trophies is "one of the oldest continuing customs in human life."[227]

> For the hunter, the display of a trophy is a reminder of the hunt and a way of extending the appreciation of the experience and the animal.
>
> —Jim Posewitz, *Beyond Fair Chase*

The trophy can make hunters selective. The deer hunter who wants the best trophy is not only looking for the animal with great genetics and good food, but he is also looking for the buck who is nearly at the end of his useful breeding lifespan, and so, nearly at the end of his life. Most game management regulations require the hunter to select the older male animals (the elk with five or more points per side or the sheep with a full curl).

The modern hunter uses the meat for nourishment and saves the most beautiful and admired inedible parts of the animal as a trophy. The trophy is a memorial to the animal and of the hunt. The trophy belongs in a place of honor, displayed in the hunter's home to show respect for the animal who died and reverence for the hunting experience. It captures the memory of the hunt and, through the telling of the hunting story, gives continuing life to the animal. It is a token of the sacrament, the hunter's communion with his animal brothers and sisters. In its presence, modern hunters are in the company of their hunting ancestors.

> A plains Indian wearing a bison skull in a ceremony may represent the buffalo in a rite simulating desired events; but he is evoking qualities and dimensions of life that the buffalo "knows" best, but which can be shared with man.
>
> —Paul Shepard, *Thinking Animals*

There has been controversy for decades about so-called trophy hunters and the conservation of game in Africa. (This also applies to other continents and countries, but Africa is probably the most well-known for trophy hunting.)

Generally speaking, here's how it works. In the absence of legalized game hunting, many wild animals are killed legally by farmers, ranchers, and villagers, as well as illegally by commercial poachers. When there is legal game hunting, the tribespeople see the wild animals as having value, and they are more likely to help protect them. When a trophy animal is killed, the hunter pays a large fee, which goes toward employment and conservation. In addition, while the hunter can ship the hides and horns back home, all of the meat is consumed locally (by workers on a conservancy or residents on tribal lands, as well as being served in the dining room at the hunting property to both the hunters and to guests who are there on photo safari).

Mounting evidence shows that the countries that allow hunting are better able to ensure a healthy population of game animals. In fact, as former USDA Forest Service Chief Jack Ward Thomas once remarked: "If you want to do a species a favor, get it on the hunted list."[228]

As we discussed earlier, selective trophy hunting can and should be used as a tool of conservation. In Namibia, conservancy-based hunting was implemented in 1990, when the country achieved independence from South Africa. Prior to independence, poaching was common, and the poacher was admired for getting meat for his family. Nowadays the animals are assets. The tribe (or the workers on a nontribal conservancy) value them, and any illegal taking of game is considered a major offense, a theft from the community as a whole.

Glen Martin writes:

> The African wildlife crisis is a crisis of misperception. Conservation has been subsumed by animal rights. These are not, however, the same things. Individual animals—most recently Cecil and Jericho—have become more important in the Age of Social Media than species stability, habitat preservation, and pragmatic if uncomfortable policies that would actually encourage the preservation of wildlife.[229]

We read in newspaper reports at the time that the local people, when asked about the death of Cecil, responded, "Who is Cecil?" The *persona* of this lion and the naming of him was an invention of special interest groups in the West.

> This is understandable: It's easier to scream in outrage over the killing of a highly charismatic lion with a cute name, sign a Change.org petition, and move on to posting selfies, than it is to actually investigate the deep forces behind the African wildlife holocaust. But emoting over Cecil isn't going to save the African lion. The African lion is not the Lion King...We don't live in a cartoon, and our problems are not solved by anthropomorphizing wildlife. Blanket trophy hunting bans may make us feel better, but they will only accelerate the slaughter.[230]

In my experience, the posing of the animal designed to reveal the physical qualities that the hunter most admires, for example, its size, the quality of its horns or antlers, the size of teeth or paws, or its general beauty...for this animal is now a particular and a personal animal: it is the animal of the hunter...The hunting photograph is the record of a relationship that will be recounted in great detail later...The photograph marks the end of the fleshy, hunted animal and the beginning of the process of re-creating and re-enlivening it in a cultural form.

—Garry Marvin, "Living with Dead Animals? Trophies as Souvenirs of the Hunt" in *Hunting—Philosophy for Everyone*

Final Denial

Throughout this book, we have emphasized humanity's connection to our roots and our kinship to the wild around us. Our marvelous human brain not only contains much of the neural circuitry of the animal kingdom, but its workings were also developed by our interactions with animals throughout our natural history. Throughout the history of our species, it has been so, and it is only very recently that this connection has been put at risk. Since our brains were formed (or built or created) by and for interaction with animals, *what can we now expect when we no longer have that connection? Will our brains atrophy? Will our empathy and bond with nature degenerate? Will we become less capable? Will we become less human?*

If hunting dies away, will all our wildness and our wilderness be next? Is this an inevitable process that turns every part of the earth into manicured parks and carefully managed zoos? Must wilderness become only a sad museum or a mausoleum that no longer offends modern mankind with nature's harsh reality? Will modern man create a world where "Bambi" will never again be eaten by wolves while still kicking and bleating in pain? Where "Bugs Bunny" will never die the horrible, painful death that awaits those who die of old age in nature?

We were whole animals before we became modern humans. Our whole way of thinking is shaped by our ancestors' way of hunting; thinking is hunting.

> Eyes to front, born to hunt,
> Eyes to side, born to hide.
>
> —Lee Shaykhet, Law Enforcement Trainer

The Politics of Hunting

Hunters prefer to deny the politics of it. In fact, there is a persistent trend among hunters to dislike politics and regard it with disdain. However, if we look at hunting as a political activity, we might gain an important insight into how others view hunting.

First and foremost, going hunting is a political statement. Hunters are champions for independence and self-reliance...not a popular concept these days.

Hunters see their rights as not only inalienable but also as bred in the bone...not as a structured liberty in which the authorities dole out each right and entitlement. Hunters tend to look at hunting as a natural part of their God-given rights, as the Declaration of Independence says: "The Laws of Nature and of Nature's God entitle them...We hold these truths to be self-evident."

Hunters voluntarily support game management and provide good outcomes for animals and for many destitute people around the world...not a generalized tax-and-spend narrative.

Hunters see hunting as an act of self-sufficiency...not a victim-oppressor equation.

Hunters see themselves as conservators of heritage and traditional values, preserving the traditions of our ancestors and, by doing that, preserving the ethical foundation of humanity.

In totemic culture, animals have neither duties nor rights, but in class society the prevailing political web envelops them, too.

—Paul Shepard, *Thinking Animals*

Women of the Wild

Coauthor Linda can definitively affirm that women who hunt are an asset to the hunting community. And hunting is an outstanding skill for women to develop. She grew up in an area where the whole family went to the hunt camp for the deer and moose seasons. In some families, the men took the dawn and dusk shifts to hunt for big game while the women took the midday shift to hunt for partridges. In other families, there was no specific gender distinction. Big game hunting required fitness and patience, while bird hunting took patience and fitness.

Women hunt for the same reason men hunt: the thrill, the trophy, the meat, the self-sufficiency, the nutrition, the fellowship of other hunters. Feminist, educationalist, and philosopher Nel Noddings says, "Bloody jaws belong to the same creature who licks her young with such affection."[231]

It's like lipstick. Although recorded in history for several thousand years (and in modern times, usually regarded as having a sexual connotation), when we look at bright red lips indicating the blood of a fresh kill, we authors think "carnivore." A related fashion is long, painted fingernails. Some people think long nails demonstrate good health, while others believe that it's a sign of a woman of leisure. But when we see bright red, exaggerated fingernails, we think "carnivore." To us authors, the symbolism is one of "fangs and claws"…a vivid picture of a successful carnivore mid-meal.

Nature writer Ruth Rudner put it this way: "What has the ease of buying food done to our awe of the animals that feed us? How awed is anyone by a cow? How many people, for that matter, cutting into a piece of cow, remember its life or have much interest in ingesting its spirit? How many ask its forgiveness?…Most women who hunt—and their numbers are steadily increasing—hunt with reverence."[232]

Mary Zeiss Stange is professor of women's studies and religion. Her book *Woman the Hunter* has been described as "the first cultural history of the relationship of women and hunting" and gave her "international recognition as the primary scholar working on the subject today."[233] As Stange says, "How well we can claim to know nature…is a complex question [that] involves looking at the nonhuman world and our relation to it, as well as looking within. Artemis knows the way, but the only way she will take us there is as a hunter."[234] Stange quotes poet Carol Frost:

> I stand in violence, in death,
> and I am happy…
> The light withdraws; chills me; alters
> nothing.
> …
> But this once
> let me tell the truth
> that can't be told
> outright. I had no pity.
> The deer's last breath
> crawled out like a clear beautiful ray
> of sun on stones. I kissed
> its head. I couldn't help myself.[235]

"That is Woman the Hunter speaking," Stange says. "Her message resonates with the Native American sensibilities…and with what we can glean of hunter-gatherer worldviews. But we need not look 'outside' Western culture to find her. Indeed, in the figure of the goddess Artemis, she stands at the dawn of Western civilization."[236]

Yet, women who post pictures of themselves in hunting situations face enormous pressure on social media. The negativity is astonishing. The study of psychology would tell us that it is most likely an indicator of great insecurity that antihunters would overreact like that.

The number of women hunters in North America has been increasing, some would say skyrocketing. The US Fish and Wildlife Service says their numbers were up 35 percent from 1991 to 2011,[237] but more recent numbers show a steeper curve. As the NRA reports: "in 2001 there were 1.8 million registered female hunters in the US, but by 2013 that number almost doubled to 3.3 million. That's an 85 percent increase in the number of female hunters nationwide."[238] According to the National Shooting Sports Foundation, that means about 20 percent of all US hunters are women.[239]

The trend is similarly strong in Canada. The number of women hunters in Alberta almost doubled in the decade prior to 2016, while in Ontario it rose 70 percent in the four years between 2012 and 2016. Similar growth rates are reflected in British Columbia and Saskatchewan.[240]

It's more difficult to get statistics for Europe and the rest of the world, but estimates indicate there are a total of 6.7 million hunters in Europe (as of 2013). Estimates of female hunting participation used to be about 1 to 4 percent. But in Norway (as of 2013), about 12 percent of hunters are female, "an increase of 60 percent over a decade."[241]

The trends are clear. Women are joining the hunt. And they're doing it in large numbers. And they're doing it all over the world.

The most recent new hunter that Linda introduced to deer hunting was a forty-year-old woman who had never hunted, didn't come from a hunting family, and whose husband had no interest in hunting. She wanted to know what it was all about, and she especially wanted to know if she could do it. She was an accomplished shooter (rifle, pistol, shotgun, and crossbow), so there was no question she would be able to shoot accurately. Her concern was being able to summon her spirit to take a life. She was able. She learned about focusing intensely, about letting the adrenaline sweep through her, about giving thanks, about field dressing the animal, about getting the meat to the butcher and to the table. She was thrilled and grateful to join the hunt as a full member.

Oh, Yes...We Use Guns

> Gunpowder was the ultimate "roar,"
> since it had both a bark and a bite.
>
> — Lt. Col. Dave Grossman, *On Combat*

Hunters use many tools to hunt, and one of them is the firearm. So, gun hunters are kind of hit by a double whammy. They are often called to defend hunting and to defend using guns.

And of course, female gun hunters are hit with an extra whammy! Stange says it best: "Perish the thought that women might take up arms, become skilled in their use, and become thereby simultaneously able to defend themselves and to fend for themselves! Woman the Hunter...is a profoundly unsettling figure, her wildness a force to be reckoned with."[242]

There are many people throughout history who well understood the power of the gun. While the hunter is mostly interested in the firearm's ability to ensure a quick, clean kill, the fact is that the gun represents one of the most political possessions a person can have. As gun owners well know, gun control is not about guns; it's about control.

> That rifle on the wall of the labourer's cottage
> or working-class flat is the symbol of democracy.
> It is our job to see that it stays there.
>
> —George Orwell

Chapter Fifteen—Summary

We are predators! As a puppy naturally chases a ball, so are we instinctively and irresistibly drawn to *la chasse*. Like a raptor detecting the distant motion of its tiny prey, so are we drawn to unexpected movement in our field of view. The hound sniffs the ground for the scent of prey, and so do we judge our surroundings by the smell in the air. And we do so *love* to follow a path that leads over the next hill to new hunting grounds and fresh opportunities.

The dawn of agriculture was also the dawn of hunting's decline, and today the pushback from antihunters can be intense. But that antagonism betrays a deep disconnect from the reality of the world. Hunting is the only way to truly understand and embrace the natural world instead of absorbing a sanitized ideal of nature from various media enabled by degrees of separation from nature itself. Hunters honestly acknowledge that for us to live, other beings must die. This is true of every human on earth, no matter their diet. And hunters consciously face the reality that they themselves will one day die as well. Hunting is true connection to the land, it is honesty, and it is our birthright.

Chapter Sixteen

We Are the Champions...of the Wild

> [T]he sportsman puts his money where his mouth is. He
> buys wetlands...pays off lumber companies to initiate better
> timbering methods; bankrolls subsidies for farmers and
> ranchers to maintain and improve wildlife habitat degraded
> by poor agricultural practices.
>
> —Jim Fergus, "From a Hunter's Road" in *A Hunter's Heart*

The wild places need the sportsman's conservation work. Each
animal needs its own type of wild area. To be clear, there's no use
trying to maintain a duck population in a desert. If all we leave
for the wildlife is marginal land, we know we're not leaving them
enough. Marginal land is not capable of sustaining much life.
We'll soon be describing the animal populations as "square miles
per animal" instead of "animals per square mile."

The Economics of Hunting
Economics is about making choices. It is about human behavior. It is
about deciding what we value and what we think it might be worth.

What Do We Value?

So-called recreational hunters famously pay a great deal for the privilege of going hunting. They rationalize the expense against the meat they bring home. But the truth is, they still go hunting when wild game meat is significantly more expensive than supermarket meat. They still hunt when wild game is more expensive than farmed game meat. They still hunt when they can't keep the meat for themselves. (This is true of safari meat, where the hunter gets a taste and the rest goes to the locals.)

As Paul Shepard says, "In the long run, animals are competitive with man. They occupy his space; they eat what he eats. They use the planet's limited energy and materials."[243] But now, our population consumes a large share of the planet's bounty, and "we know that in *our* world nothing is free. In the future every animal on Earth will exist at some cost to humanity."[244]

In an essay called "Restoring the Older Knowledge," Ted Kerasote quantifies the economics of a hunted animal. He says that a locally hunted elk "incurs a cost to planet Earth of about eighty thousand kilocalories."[245] He includes the cost of the hunter's weapon, clothing, transportation, as well as the cost of freezing the meat for a year.

> If the hunter chooses to replace the amount of calories he gets from 150 pounds of elk meat with rice and beans grown in California, the cost to Earth is nearly five hundred thousand kilocalories.[246]

He includes the costs of farming, such as irrigation and transportation.

> It does not include the cost to wildlife—songbirds, reptiles, and small mammals—killed as a by-product of agribusiness. Their deaths make the consumer of

agribusiness foods a participant in the cull of wildlife
to feed humans.[247]

Clearly, a hunted elk takes far fewer resources from the
earth than producing equivalent nutrition in rice and beans
(eighty thousand Kcal versus five hundred thousand Kcal; about
16 percent). And we gain the intrinsic value in preserving the wilderness. Raising "rice and beans" is the sole use of that acre, but
the hunted elk is a bonus. That elk's land creates multiple different
uses for multiple users. The users might include the bird-watcher,
the hiker, the camper, and the photographer. The ecotourist or the
local population may benefit. And certainly, game management
and hunting activities improve. The ecosystem improves, along
with the habitat of every other creature therein.

But wilderness areas are not as "productive" per acre. So, if
hunting is not economical, then should the whole earth be converted into the most efficient form of food production? Should
the wilderness give way to the needs of mankind, except for some
zoos and parks?

Of course not. It is not all about producing food. Nevertheless,
the cold laws of economics (like the harsh laws of nature) tell us
that someone must pay for the wilderness. Someone must make it
economically viable.

The intrinsic value of an animal, a wilderness, an experience…are all priceless. That's hard to put on a balance sheet. That's
why most hunters willingly contribute to conservation and game
management, above and beyond the hard costs of the hunt. But
when it comes to whether animals and their habitats can survive,
it's all about the hard costs. It is, therefore, all about the dollar
value of the hunt. We were impressed with Ivan Carter's attempt to
educate at least one photographic hunter by making this awesome
offer (as reported by *Sporting Classics Daily* online magazine):

Carter posted a status update...offering to take anyone interested on a safari for Cape buffalo, complete
with the up-close stalks that hunters routinely make.
The participants would close to within ethical shooting distances, set up, and "shoot" a buffalo—with a
camera.

In return, they would pay the same amount a
hunter would expect for a similar hunt: $3,000 for
the trophy fee, roughly $1,250 per day for the actual
trip, plus the anti-poaching and community fees that
are required.

In total, the photo safari would cost between
$20,000 and $25,000.[248]

Not only was Ivan Carter beautifully describing what hunters routinely do, but he also was challenging photographic hunters
to ante up.

Jim Shockey is a Canadian outdoor writer, a professional big
game outfitter, and television producer and host for many hunting shows. *Outdoor Life* magazine called him "modern hunting's
most influential celebrity."[249] In an interview on *Wide Open Spaces*
online magazine,[250] journalist David Smith reports Shockey's
comments:

"The reality on the ground in Africa is that if you
take away the value, the legal value, of these animals
it does them no favors," he said. "It's called sustainable harvest. Hunting pays for many things that
protect elephants: anti-poaching units for example.
Hunting places a value, a monetary value on these
animals."

Shockey points out that if the animals don't have a monetary value, there'll be no control. With selective hunting, the animals have a value, and specific animals will be targeted. It may be an old animal that can't fend for itself (like the warthog Keith took in Namibia) or one that can't father any youngsters but keeps the young bulls from mating (like Keith's oryx cull described later in chapter 16). Sometimes the target is a rogue animal that's dangerous to the local community.

As for elephants, Shockey says poachers tend to prefer younger animals because it's easier to carry and smuggle small tusks. "I believe that the only hope these animals have is legal hunting. But people with good intentions are ignorant of the facts on the ground."

Later in the interview, Shockey says, "They have biologists over there—smart people—who study these things and who support sustainable hunting. But you can easily ban hunting. You can't ban poaching."

Poaching

A long time ago, there was nothing called poaching because anyone could take any animal at any time. As noted in an earlier chapter, poaching arose when animals became "owned" by royalty or government. When commercial hunting (whether for meat or for tusks) started to escalate, the animal numbers plummeted, at least in some areas. That led to government-run game management, and poaching became a criminal offense. It's hard to manage a species if thieves can walk off with the inventory.

There are a lot of costs involved in protecting wild game. There are even greater costs in developing their habitat so they can flourish. Sometimes people think kind-hearted donations and photo safaris can raise enough funds. You'll remember we earlier described Ivan Carter's offer of a simulated hunting safari to anyone who wanted to take photos of animals rather than shoot them.

He had no takers. More recently, Ivan Carter has offered "Hunt and Collar" experiences. These are where the hunter hunts a plentiful animal like Cape buffalo and then darts a locally endangered animal (like the elephant) that needs collaring or medical care. The hunter then gets his picture taken with the endangered animal before it wakes up. These combination hunts are sometimes available at a fixed price or auctioned at a convention. The funds go to habitat development and anti-poaching efforts.

In the Jim Shockey interview previously quoted, Jim also talks about poaching. He says, "Somewhere along the way, hunters and poachers became kind of the same in the public eye. But they are very different." He provides an analogy: "It's like if you go to a bank and make a deposit, you're using the bank. And then after you a bank robber comes in and robs the bank. He's also using the bank. But you're both using it for very different reasons, one legally and ethically, the other illegally and unethically."[251]

Many people in Africa are working to stop bush meat poaching. But they are a small part of a much bigger economy: the illegal trafficking of tusks and horns and other body parts. An anti-poaching donation may be $100, and a dangerous-game safari hunter's trophy fee may be $40,000, and a selective hunt may bring in $350,000 or more at auction. But all these pale in comparison to a single bust of illegal rhino tusks. One such incident yielded eighteen horns with an estimated street value of more than $3.5 million. AP News (Johannesburg) reported that this was the fourth bust at O. R. Tambo International Airport between July 2020 and February 2021—in total, more than $15 million worth of horn was confiscated.[252]

Aside from habitat destruction, the main challenge Africa faces is the fight against poachers. Poaching is the biggest threat to endangered species. All of the game in Africa is protected by conservation management, but endangered species got that way

294 → On Hunting

because of illegal killing. If it were only the occasional starving family looking to put meat on the table, it might not have a big impact. But that's not the case. It's poachers looking to sell the animals (as bush meat) or a small part of them (such as tusks). They are well-armed, well-organized, and sometimes protected by tribes, rogue government bureaucracies, or international gangs. They kill animals indiscriminately and in large numbers.

Ivan Carter is one of the most outspoken members of a rising anti-poaching movement. He works tirelessly to raise the visibility of the issue, to raise funds to develop and implement solutions, and to make a difference to the animals and the people of Africa.

In India, there's a different solution. The endangered animals are protected in national parks, and the anti-poaching units are paid for by the government.

> In Kaziranga, a national park in north-eastern India, rangers shoot people to protect rhinos. The park's aggressive policing is, of course, controversial, but the results are clear: despite rising demand for illegal rhino horn, and plummeting numbers throughout Africa and southeast Asia, rhinos in Kaziranga are flourishing.[253]

In 2015, park rangers reportedly shot and killed twenty poachers.

What Hunters Contribute to the Economy

Hunters contribute to the economies of many countries. North American hunters make big contributions to the hunting economy. In the US, sportsmen contribute nearly $9.4 million *daily* toward conservation through licenses, excise taxes, and other special taxes such as duck stamps. That equates to over $3.4 billion annually for conservation.[254] And still, as Frances Stead Sellers of the *Washington Post* reported in 2020, "Hunting is 'slowly dying off,' and that has created a crisis for the nation's many endangered species."[255] Canada is a relatively small country, but about 3.5% of the population hunt, and hunting generated a $4.1 billion contribution to Canada's GDP in 2018.[256]

European hunting raises considerable funds to support hunting, game management, conservation, and habitat protection. As reported by FACE (European Federation of Associations for Hunting and Conservation) at a conference in September 2016, the numbers are amazing. In addition to the direct economic contribution (16 billion euros), it's estimated that the indirect contribution provides an additional estimated €16 billion.[257]

It is somewhat difficult to get numbers from South America, but in his PhD dissertation, William C. Gregory says that hunting can help. Hunting provides economic advantage to the poorer, more remote areas that have fewer ways to generate income for their residents.[258] It is even more difficult to get financial information on hunting in Asia. However, the existence of such hunting consultant companies as The Hunting Consortium, Ltd. is a good indication that there is a hunting industry alive and well in these areas.

In Africa, the numbers (when they are available) are country by country. The best-known countries for hunting are South Africa, Zimbabwe, Tanzania, and Namibia. Namibia is the poster child for sustainable hunting, with 42 percent of their land now under conservation management. Even the World Wildlife Fund sings their praises:

> Namibia was the first African country to incorporate protection of the environment into its constitution, and the government gave people living in communal areas the opportunity to manage their natural resources through the creation of communal conservancies. These conservancies—as well as governments,

nonprofit organizations, and other entities—have restored populations of lions, cheetahs, black rhinos, zebras, and other native wildlife to the world's richest dry land.[259]

In many areas of Africa, years of pastoral grazing has savaged the land. Conservation funding, raised by hunting, goes toward reclaiming land. To ensure the success of this endeavor, the local people must support it. And in order to support it, the local people must be able to earn a living and feed their families as well as they could by herding cattle.

> There is no doubt that hunters spend more time on the ground interacting with as well as supporting local communities than any other group of tourists.
>
> —The Hon. Uahekua Herunga MP, Minister of Environment and Tourism, Namibia

In the jungles of Africa, hunting is a tool that gives conservation value to the rainforest. Commercial stripping desecrates the rainforest. This results in decreasing wildlife habitat. And this means increasing poverty for local Indigenous tribes. Safari hunting has helped provide nonintrusive land use. The safaris increase the value of the jungle to the locals. The locals then want to conserve and manage the rainforest. We recently watched a show about a jungle safari in Liberia. Not for the faint of heart! Professional guide and hunter Rob Dunham worked and sweated through the dense jungle. It was a spot-and-stalk hunt through heavy undergrowth. In truth, it was more like glimpse-and-slog hunt. Even the scholars are taking note. They generally conclude that mixed-use rainforest endeavors that include the locals are more likely to achieve the desired conservation goals.

In the US, the conservation funding raised by hunters has had an outstanding effect. The National Shooting Sports Foundation reports:

White-tailed Deer Population	In 1900: 500,000
	Today: 32,000,000
Ducks/Waterfowl Population	In 1901: Few
	Today: 44,000,000
Rocky Mountain Elk	In 1907: 41,000
	Today: 1,000,000
Wild Turkeys	In 1900: 100,000
	Today: 7,000,000
Pronghorn Antelope	In the 1960s: 12,000
	Today: 1,100,000[260]

> In a civilized and cultivated country, wild animals only continue to exist at all when preserved by sportsmen.
>
> —Theodore Roosevelt

Values of Trophy Tags: Draws and Auctions

Lottery draws are a common means to distribute a small number of desirable game tags to many potential hunters. For example, in some parts of Ontario, Canada, a lottery system allocates bull moose tags and antlerless deer tags. In North America, lotteries allot tags for wild sheep. Many people have entered the lottery annually for a lifetime and never had a chance to go sheep hunting.

Auctions also provide a distribution method. This is an important source of funding for conservation. Sheep tags often fetch hundreds of thousands of dollars *each*. The top price ever paid (as of this writing) was $480,000. The 2016 auction at the

Wild Sheep Foundation raised over $3,000,000, and the foundation added to that, contributing $4,700,000 to conservation. The money goes to state or provincial game departments or directly to Indian reservations. Conservation donations also come from individuals, such as the hunter who paid $3,000,000 for fencing to protect wild sheep from highway traffic.[261]

African auctions also raise money for conservation. In the past decade this has been a source of controversy, and the auction prices have dropped significantly. This loss of income has harmed the conservation efforts in many countries. However, selected animals are still available for hunts, and auctioned tags fetch hundreds of thousands of dollars each.

Many ecologists are now supporting and promoting selective hunting. While the awareness of the general public lags behind the subject-matter experts, more and more media stories acknowledge that hunting is not the problem. In fact, hunting is part of the solution. Many people (especially hunters) are working to increase animal populations through habitat protection and rewilding, anti-poaching programs and helping ranchers protect their domestic animals.

The Conservation Movement

As reported by the CBC in a documentary film called *Angry Inuk*, which released on May 2, 2016:

> Animal rights groups have been fighting since
> the 1960s to shut down the sealskin trade. Using
> graphic campaigns featuring fluffy seal pups being
> bludgeoned by hunters, opposition to seal hunting
> mounted. In 1983, Greenpeace and other groups
> were finally successful in getting the European Union
> to ban sealskin products made from white coat harp
> seal pups.

Even though the legislation targeted only one kind of sealskin, the campaigners ruined the reputation for all types of sealskin.[262]

This and other reports describe downstream results as follows:

- At the time, little thought was given to the impact the ban would have on the Inuit.
- Although the Inuit were exempt from the ban, the market for sealskin evaporated.
- A year later, the average income of an Inuit seal hunter in Resolute Bay fell from fifty-three thousand dollars to one thousand dollars.
- Suicide rates were already climbing in Inuit communities and spiked to become the highest in the world.
- Two generations later, seven in ten Inuit children go to school hungry.
- Currently, Inuit communities have the highest poverty and unemployment rates in North America, housing costs are among the highest in Canada, and food prices are probably the highest in North America.

Finally, in 2014, Greenpeace openly apologized to the Inuit people of North America and Greenland for its role in causing them forty years of grief, hardship, and frustration.

Conservation and Preservation
Across the hunting community, there seems to be a powerful consensus about conservation. Hunters understand that habitat is the base requirement for healthy, diverse, self-willed wildlife. Hunters have been working toward reclaiming and improving habitat for decades. The hunting community is now well into the second century of doing so, persistently, consistently, and very successfully. The traditional conservationists lost the support of hunters when

they wanted to preserve wild areas with no people. However, they are now talking about the ideas of rewilding, wilderness areas, connectivity corridors, and the inclusion of predators as valuable members of the wild community. This is a point of view that hunters can support.

The hunting community has come to believe that the practicality of the new conservationists is useful. This pragmatism can help to ensure that human projects have three important outcomes. They can minimize their immediate damage to habitats. They can provide high-value wildlife areas. And they can encourage returning land to a wild state. This is not new to hunters (they have been doing it at least since 1871 and the Native Americans for thousands of years previously), but at least it's common ground and one idea everyone can support.

> Dr. Paul Shepard, the father of modern (deep) ecology...
> ironically becoming the guru for the urban ecologist, the
> modern conservationists and the Green movement...
> advocated for a genuine understanding of man the hunter,
> and a return to our Pleistocene roots. Perhaps we can find a
> commonality through his wisdom.
>
> —David Petersen, *Heartsblood*

If hunting is to continue to exist, then wildlands are an important part of that. Modern hunters have always had to exist in both the civilized and the wild worlds. Perhaps it is that grounding, that wild wisdom, that's allowed the hunting community to be the true world leaders in conservation.

The larger questions of conservation can be difficult to latch on to. The complexities can be overwhelming. But the good news is that Mom was right. It's better to clean up your own corner of the universe than to worry about everyone else's. As Terry Tempest Williams wrote:

It is only through the power of our own encounters and explorations of the wild that we can cultivate hope because we have experienced both the awe and humility in nature. We can passionately enter into the politics of place, even the realm of public policy, and change it, if we dare to speak from the authority of our residences.[263]

By sticking to what we know and where we know it, we can speak with passion, authority, insight, and credibility.

Game Management

> We call it "game management," but it's also "people management." And with 7 billion of us—*the vast majority of whom don't care*—it's getting to be quite a big management issue.
>
> —Alex Robinson and Natalie Krebs, *Outdoor Life online magazine*

Game management is a tool used to ensure that the optimum mix of wildlife in each environment can flourish. Hunters are the primary instrument for game management worldwide. Hunting is a tool to fund wildlife projects. Hunting is also a tool to manage the carrying load of each ecosystem.

Hunters are one of the biggest volunteer forces in North America. Think of it this way: hunters pay for licenses so they can go out into the bush and make improvements for wildlife. While they're out there, they help maintain the target levels of wildlife according to government standards.

When Linda and Keith hunted in Namibia the second time, they shot some cull animals. Culling is also an important part of game management. The culls were selective. There was one old oryx who ruled several square kilometers of the Kalahari. He was not able to father any more babies, but he was able to fight off any other bulls who wanted to. The landowner had not been able to cull him. The oryx had a "comfort zone" of well over four hundred meters, and that made an ethical shot very difficult. Keith had the equipment and the skill to ensure a clean kill, and he did so. It was more demanding than most trophy shots, and it was as exciting as any hunt. The meat all went into the cooler to feed the staff. Because Keith and Linda respect every animal they kill, they had the horns mounted, and the oryx has a place of honor in their home. They treat all their game animals the same way: ethical shots, meat eaten, and memory honored.

There is a tale to every trophy. Keith and Linda remember every one of these noble game animals and maybe even immortalize them in a story or with a picture. And there will always be something special about a cull taken from 485 meters (530 yards) an ocean away from home. Not every hunter can have the opportunity to go to Namibia to shoot cull animals. But worldwide, it is responsible, conscientious, ethical hunters who are all taking care of game management in their little corners of the universe.

What Would It Be Like without Hunting?

In *A Hunter's Heart*, author and hunter David Petersen interviews his friend and fellow hunter, Dan. Referring to fringe animal rights activists, who would, if they could, create a world without hunting (without meat-eaters, without predation), Dan says:

> More often than not, their arguments are biologically naïve, unworkable, and ultimately immoral…What the fanatical animal-rights types refuse to acknowledge is that humans evolved as predators, and nature needs predation, especially these days. Hunters didn't design the system, and now, with most of the big natural predators wiped out to make the world safe for our domesticated livestock, hunters, as ironic as it

may sound to the uninformed, have become wildlife's closest allies.[264]

It's far too late for humans to abandon game management. We therefore cannot (and must not) abandon hunting as a management tool. In North America, a common wildlife issue arises with deer populations. Suburbanites generally love seeing "Bambi" in their neighborhood. Eventually, the local deer population explodes from lack of predation and excessive restraints on hunters. Then the deer move into the suburban yards and eat the daylilies. On the nearby roadways, they are coming through the windshields of mini-vans and SUVs traveling at highway speeds.

Coauthor Dave rode with a Pennsylvania State Trooper on a stretch of interstate highway. Mile after mile, you were never out of sight of the carcass of a road-kill deer. Each one represented not only the unfortunate death of wildlife but also large amounts of money to repair damage to a vehicle. On occasion, that carcass was mute testimony to the death or injury of citizens as the deer came through a windshield or wiped out a motorcyclist. Suddenly Bambi looks a little different, and the locals begin to understand the basic concept of game management.

Overpopulations of deer reduce biodiversity by nipping every little tender sprout from the ground to their five-foot browse line. Nearby greenbelts (bird sanctuaries, parks, wooded stream areas) provide corridors for deer to move into suburbia, at least when they want to feed. These animals are contributing to the spread of disease to both humans and their pets. In addition to Lyme disease, which can infect humans, there are other serious deer diseases that can pass to elk and moose as well as pets.

Overpopulation can happen quickly. Each doe reaches sexual maturity in one year, and they almost all breed, often bearing twins and sometimes triplets. Under optimal conditions, the deer population, if unchecked, doubles about every two years. In the latest numbers available, there are over thirty million deer in the US. (Hunting currently takes six million deer, and with an estimated twelve million fawns born each year, the doubling takes a little longer than two years.)[265]

In the face of environmental depredation, the idea of human nonintervention, of "letting nature be nature," is therefore seductive. It is also resolutely unrealistic.

—Mary Zeiss Stange, *Woman the Hunter*

- With hunting, wild animals have value. Hunters value the wild animals.
  › Without hunting, wild animals become pests. Pests will be killed until they are eradicated.

- With hunting, wildland has value. Hunters value the wildland.
  › Without hunting, wildland has few or no champions. Without hunting, the wildland will disappear under the expansion of herding, ranching, agriculture, and towns. Both farms and towns increase the animal-human conflicts. More animals will lose this battle. More animals will die.

- With hunting, conservation efforts are well-funded.
  › Without hunting, conservation loses a major source of funding. Without conservation funding, habitats disappear, and habitat corridors are isolated. Without hunting, wild animals have no place to be and no place to go.

- With hunting, poaching is a crime, and anti-poaching efforts receive funding.

> › Without hunting, anti-poaching units lose an important source of funding and often evaporate, and poachers can take whatever they want.

- With hunting, game management is an effective way to ensure many species survive and thrive.
  > › Without hunting, game management loses funding and loses the largest volunteer workforce to support wildlife and their habitat.

> The notion that nature can and will take care of itself is a truism that anti-hunting, anti-wildlife management advocates often declare. However, nature often chooses to take care of itself in ways that we may not feel happy about at all. Nature has used extinction as a management tool far more than humans have ever been responsible for. I doubt any of us would be okay with the extinction of lions.
>
> —Merelize van der Merwe, in a *Facebook* post (February 27, 2016) on "The Cecil Effect"

What Is Selective Hunting?

In the *Wide Open Spaces* interview we previously quoted, Jim Shockey says:

> The fact is that trophy hunters are selective hunters. We select specific animals, for a number of reasons, and remove that single animal from the herd, for the well-being of the herd as a whole.

Shockey says the term *trophy hunting* is now derogatory. He thinks it began possibly in the early 1960s. The meaning became dirty. The "big white hunter" posed for pictures, looking like the conquering hero. Disney's *Bambi* villainized hunters. In fact, the humans in the movie were poachers, not hunters. But that's not what the movie-goers thought.

> Somewhere along the way, the image of the hunter began to become that of someone who cuts the head off an animal and leaves the rest. But that never happens. That's a fiction. Only poachers do that sort of thing. And hunters aren't poachers.[266]

In 2017, *The New York Times* ran a story with the tagline "Permits to hunt bighorn sheep are auctioned for hundreds of thousands of dollars—and that money has helped revive wild sheep populations and expand their territory."

Just to show how very selective a hunt can be, here's the story. There's a particular sheep that the hunters have been watching, possibly a state record. The hunters and the outfitters know about this one animal.

> That drove the price up. But he's old. He's 11 or 12. The question is whether he'll still be alive, if he'll make it through this winter. And then the question is whether we can find him.[267]

Whether the hunter who gets the tag finds that ram in the designated hunting area is uncertain. During the rut, the rams have a large territory, moving around to find receptive ewes.

As John Branch reports, "Whether he gets a ram or not, the California Department of Fish and Wildlife receives 95 percent of the auction price—in this case, a record $223,250. That feeds into a general big-game budget of about $10 million…The money pays for sheep-specific employees and conservation efforts."[268] That's a lot of money.

Balancing the Challenge

> Wild places are uncontrolled. Their presence reminds us of just
> how far we have distanced ourselves from the rest of nature.
>
> —Roderick Frazier Nash, *Keeping the Wild*

There are signs that a balance is stirring in some corners. The traditional conservation movement has made gains by respecting and cooperating with the tattered remains of our North American hunter-gatherers. In an essay in *Keeping the Wild*, Curt Meine (conservation biologist, historian, and writer) cites a couple of examples. In one modern Indigenous conservation effort, there's a partnership between the Cree Nation and Parks Canada. This project is to create a protected area along the east coast of James Bay (where hydro-electric projects abound). Another project, by a band of the Chippewa, is to establish the first ever tribal wildland park in the US. Meine suggests that perhaps these projects "can serve as reminders and guideposts, showing that concepts of *home* and *wilderness* are not, and never have been, as antithetical as we sometimes presume."[269]

In the same book, Professor Brendan Mackey emphasizes that we have two jobs. We need to "avoid introducing threats to intact landscapes and to reduce or eliminate threats to...landscapes suffering loss and degradation." He also advocates for connectivity corridors, where animals can move from one protected area to another. In Australia, this approach, called "connectivity conservation," has taken root. It is happening through a synergistic combination of grassroots and government. One project includes continental-scaled corridors.

More and more, habitat corridors are resolving habitat fragmentation.

> Wilderness is rudimentary and fundamental in ways that we've mostly lost as a culture. This loss, by the way, weakens us. Wilderness strengthens us.
>
> —Howie Wolke, "Wilderness: What and Why?" in *Keeping the Wild*

Hunting Organizations Are Part of the Solution

Many of the hunters reading this book probably already belong to at least one hunting or conservation organization, and they support it through donations and by volunteering. In North America, hunters have several of their own well-known conservation projects, such as National Wild Turkey Federation, Wild Sheep Foundation, Boone and Crockett Club, Rocky Mountain Elk Foundation, Safari Club, and many others. Some of these have grown into international organizations.

On other continents, there are such organizations as the African Wildlife Foundation, the Ivan Carter Wildlife Conservation Alliance, the International Council for Game and Wildlife Conservation (CIC), the European Federation of Associations for Hunting and Conservation (FACE), and many, many more.

> And where shall our spirits dwell when wild nature dies?
>
> —David Petersen, *Heartsblood*

There are also several worldwide conservation organizations that support hunting, though you'd never know it! Coauthor Linda once had a discussion with an antihunter who gradually revealed he had no hunting knowledge beyond what the media told him. Eventually, it became clear that he didn't want to improve his knowledge. At the end, he "proved" his patronage of nonhunting conservation by saying that he regularly donates to the World Wildlife Fund (WWF). Maybe he regularly buys the cute little stuffed toys promoted by WWF. Clearly, he had not investigated the details of the WWF projects, which include sustainable hunting. The WWF is clear:

> As such, WWF holds the position that trophy hunting is a potential conservation tool that can be considered as part of an overall conservation strategy, including for threatened species.[270]

The website goes on to describe successful projects in both Namibia and Pakistan. It's fine to buy those cute little stuffed toys. They're adorable. But it might also be good to fully understand what the WWF supports in order to foster conservation efforts worldwide. Indeed, any serious conservation organization would definitely support the positive impact that hunting can have.

Chapter Sixteen—Summary

Economics and ecology in the modern world are not just about making choices. To be truly effective and beneficial, these endeavors must be about natural human behavior and inclination. All such efforts must be about deciding what we innately, intrinsically value and how much of our precious time and treasure we are willing to invest.

Hunters have made their choice. They have spoken by the millions with their wallets and their feet. They have established and proclaimed what is of value to them by investing huge sums of money and effort, with enormous potential for even more in the future.

Hunters have invested themselves in the wild, in the wilderness. Hunters can and will sustain the economics and ecology of our planet through global conservation movements, game management across vast swaths of land, and millions of hours of volunteer service.

Hunters represent the best hope, the deepest pockets, and the greatest chance to sustain our ecology and protect wildlife and wilderness around the planet. Hunting is a natural part of us, and it is a natural force for good in our lives and in our world.

Chapter Seventeen

The Hunter's Peace

When it comes to the hunter's peace, most of us can affirm from our own personal experience: you really can find yourself by losing yourself. And there is a great body of research that affirms this simple idea. The wonderful union with the natural world is what keeps so many people going back into the woods.

Not just the hunter, but the hiker, the camper, the kayaker, the canoeist, the bird-watcher, the fisherman, the photographer, and countless others all seek the wilderness. They want the calmness of mind, the relaxed purposefulness, and the bonding with something bigger. But of all the people in the woods, at this point it has been established and proven that the ethical hunter is the one who contributes the most. We've demonstrated that the hunter is the one who pays the most, economically. It's the hunter who makes it feasible for the wilderness to be available to all the other users as well. And we've confirmed that it's the hunter who attains the deepest and most powerful relationship with nature. It's the hunter who taps into his primordial roots and becomes a part of the circle of life.

Through good weather and bad, through thick and thin, whether or not he brings an animal home to the table, the hunter rejoices in the hunt. In the tranquil days of summer, the woods resound with all manner of human recreation. But precious few are those in the woods at the crack of dawn in the bitter cold days of fall and winter. Yet this is exactly the time when hunters are in the woods. All hunters, at one time or another, ask themselves, "Why can't we hunt when the weather is good?" But they know the answer. Because the cold season is when the herd can be ethically culled while making the best contribution to game management. This is the time of year when the hunter is alone with the animals of the forests and plains. This is the time for the hunter's peace.

> I am a Canadian, and people of many other cultures experience these feelings too, for this spiritual connection is a response to the power of wild places—like mountains, forests, and rivers—and not the invention of any particular culture.
>
> —Harvey Locke, "Conservation" in *Keeping the Wild*

Each of the coauthors of this book has felt the unity of life and purpose that is the hunt. Every hunter has his or her own story. Here's coauthor Linda's story:

> I was introduced early into the wilderness. My dad was a good bushman in Canada's north. He thought it was important for a youngster to be able to handle a map and compass. I remember he would take me into the bush and expect me to be able to lead us back out. He introduced me to small game hunting, rabbit and partridge.
>
> In later years, I remember lying on the beach by myself, warmly tucked into my eiderdown sleep-

ing bag, watching the Northern Lights, fascinated and in awe.

In modern parlance, I identify as a hunter.

I'm not sure when this happened. It might have been when I was a teenager and I stalked a wolf pack…no gun in hand, no intention of harming any of them, just stalking. It might have been the first time I shot a deer, it looking as surprised as I was. It might have been that long stalk toward a moose, a moose I didn't shoot, a moose that passed within about ten feet of where I stood. It might have been a night hunt for feral pigs, when I felt the rush of the kill. It might have been when, closing in on an oryx, I lay on the warm Kalahari sands and felt the heartbeat of my ancestors.

Words Matter

For us hunting wasn't a sport. It was a way to be intimate with nature, that intimacy providing us with wild unprocessed food…We lived close to the animals we ate. We knew their habits and that knowledge deepened our thanks to them and the land that made them.

—Ted Kerasote, *Merle's Door: Lessons from a Freethinking Dog*

As we've mentioned before, the word *hunting* rarely stands alone. Most people put some sort of descriptive term ahead of it. Often, that term is scornful. We can change that terminology. As Jim Shockey says in the *Wide Open Spaces* interview, "At least we can offer a different narrative. As hunters, every time we talk to the media we can reinforce the concept of selective hunting. Because we are all selective hunters."[271]

As a hunter or a scholar (or as a huntologist, who would strive to combine the two) trying to influence this debate, you may question how often most of us "talk to the media," but many of us post on social media frequently. And just because it's a "private group" doesn't mean your comments (and your pictures) won't get more exposure than you had planned. One of our friends who read an early draft of this book pointed out that every time we talk to a friend about hunting, we are influencing the debate, and we should be using the right terms.

All hunters should ensure that they present their animals respectfully. Keith and Linda always have wet wipes with them to take the blood off the animal's face, and they (well, Linda) always tuck the tongue back in the mouth. This is respectful to the animal. That picture is going to live a long life, and it's only right to make sure the animal looks proper and peaceful in death. Even more so, consider the fact that when you eat this animal, when you consume your kill, its flesh becomes your flesh. The atoms and molecules of this noble beast will go into your body and will become part of your body. Every chef knows the value of "presentation" of food on the plate, and how much more so in this case? And then when you consider that this beast died so you can live, that it will live on to become a part of you, is that not worthy of deep respect?

Deep Roots and Bright Future
By hunting, human beings can satisfy those needs most deeply embedded in all humanity, from blood cell to brain cell. If there are some people on this earth who understand this, our civilization will be able to hold on to a bright future for hunting.

Like a Hunger for Love
Earlier in this book, we discussed the role of dopamine in hunting and the control the midbrain has over some of our fondest

activities: feeding and breeding. In a lecture on dopamine and addictions, Eric Bowman added this perspective:

> [T]he key difference between human brains and those of other animals is the size and complexity of our cerebral cortex...We therefore tend to focus our attention on this area, believing that our unique mental life is due to this masterpiece...
>
> But we often ignore the bits that are nearly identical between humans and animals, such as the tiny group of brain cells that use the chemical dopamine to communicate with other brain cells.[272]

Bowman goes on to describe the role of dopamine in many physical and mental processes, as well as its role in drug addiction. Of significance, he says that one of the most important aspects of dopamine function is learning. Dopamine helps us to distinguish between rewarding and nonrewarding activities and behaviors. He concludes with this powerful observation: "Our special cerebral cortex may be in control of our actions, but our primitive dopamine system may very well serve as its teacher."[273]

Yes, we're ruled by our hunting brain!

And you, I, we, will fight a lot harder to preserve what we love—enlightened self-interest at its natural best.

—David Petersen, *Heartsblood*

As Paul Shepard wrote in *Thinking Animals,* our brain is "made possible by primate society and predator ecology."[274] If meat-eaters don't learn to hunt, they die. Conversely, animals who live are those who have learned to hunt. In Shepard's terms, what "we achieved as hunter-primates was a means of learning."[275] The first steps of that learning occur in the midbrain, where feeding and breeding (and dopamine) cemented our behavior. The mid-brain then teaches the prefrontal brain (forebrain) what it needs.

Even without being part of the hunt, we can see ourselves as though we were. Picture your children and grandchildren squealing with delight when you chase them and catch them, or they chase and catch you. We are programmed to hunt.

And when we don't follow our genetic programming to stalk, kill, and eat, our brain sends error messages. This is particularly good news because it confirms that it's a common thread across humanity. By hunting, we can remain true to our most ancient and venerable selves. The need to hunt is innate; it is born in us and borne by us like a hunger for love.

Home Is Where the Hearth Is

Where do you feel most comfortable? Where do you feel "home"? For most hunters, it's in the great outdoors. As Edward Abbey (environmental activist and essayist) said:

> For many…the out-of-doors is our true ancestral estate. For a mere five thousand years we have grubbed in the soil and laid brick upon brick to build the cities; but for [much longer] before that we lived the leisurely, free, and adventurous life of hunters and gatherers, warriors and tamers of horses. How can we pluck that deep root of feeling from [our] consciousness? Impossible.[276]

Many of us say "home is where the heart is." We also say "home is where the hearth is." From the sundowner in Africa to the bonfire in England to the campfire in North America, the purpose of the hearth is for cooking and warmth and (in some cases) security against wild animals. The control of fire was a momentous event in the history of humanity. And in our campfires today, an ember of our ancient ancestors glows in the coals.

> Everything that lives and moves will be your food. Just as I gave you the green grasses, I now give you everything.
>
> —Genesis 9:3 CEB

Even people who've never been hunting recognize that the outdoor firepit and the indoor hearth continue to draw us around and warm our hearts. It's part of our roots. In *Heartsblood,* David Petersen writes:

> [Ortega says] hunting (and the preservation of wildness it implies) is in fact essential to the maintenance of any truly "civilized society" and thus can never become obsolete. How can we expect to nurture our continued [growth] as a species by ripping out ninety-nine percent of the roots of that [development]? Hunting and gathering…are those roots.[277]

Our Souls Are in the Wilderness

> You know, I think if people stay somewhere long enough… the spirits will begin to speak to them. It's the power of the spirits coming up from the land. The spirits and the old powers aren't lost, they just need people to be around long enough and the spirits will begin to influence them.
>
> —Gary Snyder (quoting a Crow elder), *The Practice of the Wild*

A long time ago, when everyone lived in the wilderness, the children were naturally raised in the wilderness. Sadly, the chain has been broken, and we now find many children who have never risen at dawn to watch the sun, to get their feet wet in the dewy grass, to participate in the stirring of nature as all the daytime creatures awaken and go about their business of feeding, breeding, and nesting.

In *The Practice of the Wild*, Gary Snyder writes about his experiences with the Inupiaq peoples of Alaska. He says the tribal elders come to the village schools and teach traditional culture. In an evening of singing and dancing, they note the similarities to the work of poets in Western culture.

> An Inupiaq couple who had also come in from outside
> the village for the reading commented on the antiq-
> uity of myth. Our ancestors, they said, told the same
> stories as the Greeks, and the people in India, and the
> rest of Native America. *We all had a classical culture*.[278]

We all also had sacred lands. For the Inupiaq and peoples like them, the land that sustains them with hunting and gathering is their sacred land. For newer arrivals, common land is perhaps the last remnant of wild and sacred land for the ordinary American.

There is a beauty in a subsistence lifestyle, a simplicity and honesty that is hard to find in modern life. A person who has acknowledged and embraced the idea of taking a life in order to sustain his own has achieved a freedom that can't be understood by those who are in denial. For those who buy the food that others have killed, their distance from the source of the food enables them, as Gary Snyder says, "to be superficially more comfortable, and distinctly more ignorant." He adds:

> Eating is a sacrament. The grace we say clears our
> hearts and guides the children and welcomes the
> guest, all at the same time...
>   We too will be offerings—we are all edible. And
> if we are not devoured quickly, we are big enough

(like the old down trees) to provide a long slow meal to the smaller critters.[279]

An Echo of the Rhythm of the Past

> For much of its natural history the human species has been able to feel the pulse of the Earth.
>
> —Lisa Krall, *Keeping the Wild*

In the northern hemisphere, in July, August, and September, the promise of a good harvest and a successful hunt means that food and survival are assured. These months have always been a natural time of procreation, and babies conceived in those months would be born in April, May, and June. It is no coincidence that these three months have been traditional names for baby girls.

This pattern holds true across the latitudes of the earth. In temperate latitudes of the northern hemisphere, the birth rates tend to peak in the spring, about nine months or so after the harvest. "In the southern hemisphere, the months are opposite; for example, in French Guiana, women give birth in greatest numbers in October."[280]

This rhythm goes back before agriculture, back to the hunter-gatherers who would be able to harvest nature's bounty toward the end of the growing season and store it to protect against harder times in the winter. Dried fish, meat, and berries can be easily stored, and they keep better when they are cold or frozen.

> Our interaction with the world is not divisible from it.
>
> —Alison Acton, "Getting By with a Little Help from My Hunter: Riding to Hounds in English Foxhound Packs" in *Hunting— Philosophy for Everyone*

Rewilding

One of the most promising types of conservation projects is rewilding. Rewilding is letting Mother Nature reestablish her dominance. This is happening in wilderness areas, crown/common lands, national parks and forests, conservancies, and private lands. In most cases, hunting is a tool to fund the wildlife projects and to manage the carrying load of each ecosystem. One of the most exciting rewilding projects is a nature conservancy in Zimbabwe encompassing eight hundred fifty thousand acres. Ivan Carter describes it this way on his Facebook page:

> Proud to be a part of the Bubye Valley Conservancy lion project—a project that indeed wouldn't exist without the hunting model…this is a remarkable property, which just 30 years ago was an enormous cattle ranch—today it's been completely "rewilded" and thanks to some amazing visionaries has had much of its incredible wildlife diversity restored—arguably the most successful lion project in Africa, the 850,000 acre property boasts a population in excess of 400 of these endangered predators—all from 17 individuals and a few others that have "broken in" over the years—yes indeed these are the initiatives and successes so well deserving of our support![281]

Rewilding is clearly a good way to restore wilderness and wildlife. There are many initiatives all over the world, most of them meeting with great success. It's not always easy, but with diligence and the support of people willing to put money into it, it's worth it.

We also need to rewild people. Humanity needs to learn from the few remaining Indigenous peoples, and we need to learn from hunters. In modern times, hunting wild animals is foreign to most people. There is a certain amount of "culture shock" when modern people experience the realities of the hunt for the first time. Since they have not hunted, they genuinely don't understand what the hunt is all about. They, more than any others, need to be "rewilded."

Everybody Needs Nature
Childhood is an echo of our ancient hunter-gatherer past, structuring our early learning. We know that Indigenous peoples and modern hunters are physically and mentally complete only when the wilderness envelops them. Now, studies show that urbanites also benefit from exposure to nature. Even people who don't think about nature and wildness and don't particularly think they need the wild respond better when they get some exposure to nature. In a study conducted at the University of Melbourne, Kate Lee tested urban adults' cognitive abilities and concluded that "green micro-breaks" boosted sustained attention. Sustained attention is crucial in daily life and underlies successful cognitive functioning.[282]

> Modern work drains attention throughout the day, so providing boosted "green micro-breaks" may provide mental top-ups to offset declining attention.[283]

It should come as no surprise to hunters that we all can think better when we include regular "doses" of nature and wildness in our lives.

> Outdoor people constitute a close fraternity, often international in its membership. Like music and art, love of nature is a common language that can transcend political or social boundaries.
> —Former US President Jimmy Carter, "A Childhood Outdoors" in *A Hunter's Heart*

Resilience of Nature

Rewilding speaks to the resilience of nature. If you have ever had the opportunity to watch it happen, you'll be surprised how quickly Mother Nature reclaims her own. While it may take a while to grow a great oak, regeneration of wild places starts quickly. In the temperate zone of North America, after a bush fire, you can look forward to a big crop of wild blueberries in a year or so. Then come the poplar trees, slender and crisp, providing great food for moose and deer. Before long, the beavers move in and reclaim the waterways for their own. The process continues, and gradually, as the habitat changes and develops, there are more wild critters of all sorts.

When an area is reclaimed from agriculture, the same process happens. Linda remembers trying to find her grandparents' farm about twenty or thirty years after they left it. Not only could she not find the farm, but she also couldn't travel the last two miles of the road by car. The trees had taken over. When coauthors Keith and Linda reclaimed some of the roads and trails on their range property (roads that hadn't been used for about forty years), it was very difficult to find the old pathways…nature had taken them back.

Even when nature must recover from a catastrophic event, it's amazing how quickly the animals recover. In an article entitled "Animals Rule Chernobyl Three Decades after Nuclear Disaster," John Wendle writes:

> It may seem strange that Chernobyl, an area known for the deadliest nuclear accident in history, could become a refuge for all kinds of animals—from moose, deer, beaver, and owls to more exotic species like brown bear, lynx, and wolves—but that is exactly what…scientists think has happened.[284]

In fact, there are so many animals, the area now must provide guards to prevent illegal poaching. As Dave Forman points out in his essay in *Keeping the Wild*:

> Ecosystems often can recover from human impacts over periods of time, depending on the level of impact. This resilience should never be used as justification for further intrusions into wilderness, but it does provide a valid rationale for the concept of wilderness recovery and rewilding.[285]

So, just as our own bodies heal—unless we've completely compromised a core system—Mother Nature can return to her previous healthful life.

The Importance of Wild Nature

Nature is what made us. Our future depends on nature continuing to form and shape us. We and all the other species are meant to be adaptive and resilient. Only by allowing wild nature enough space and dominion will we be able to continue our natural development.

Our natural progression may take a different path than we expect. Sometimes it's difficult to accept that nature is wild and that conservation doesn't necessarily mean preservation. In fact, for nature, wildlife, and humans to be adaptive, we must all continue to be dynamic and to evolve.

It's Worth Fighting for Natural Rewilding

What's important and what's worth fighting for is to have significant areas of our blue planet where nature is the dominant force. We humans are part of nature, to be sure, but at least some of the time, we need to back off and let nature take its course. Only by letting this happen will nature be able to evolve as intended.

There are those who believe that humans can orchestrate nature, planning and controlling conservation of plants, animals, and humans. They compare it to growing a garden, where the "good plants" are nourished and the "weeds" are pulled out. In an essay called "The Future of Conservation" (included in *Keeping the Wild*), Brendan Mackey writes:

> A conservation goal defined by the gardening met-aphor, however, takes us down the wrong path...It would replace natural selection with human deci-sion-making dominated by the desires to optimize for efficiency and maximize short-term gains.[286]

Unfortunately, the "garden" version of conservation is guided by the arrogant presumption that mankind can better manage nature's resources than nature can on her own.

The Wildness in Us

Wildness is what made us. Being resourceful and resilient enough to find food shaped us and made our bodies strong. Learning about plants and animals and finding food (hunting, fishing, gathering) formed our minds. Living in small groups and respecting each member for what they could provide molded our social nature.

Our future depends on wildness continuing to form and shape us. We and all the other species are meant to be adaptive and robust. Only by allowing wildness enough space and domin-ion will we be able to continue on our natural path. Only by immersing ourselves in wildness can we let the wildness live in us.

Many people, including some hunters, are uneasy about predators. Sometimes they simply fear predators, and sometimes they transfer that fear to those who hunt predators. But hunting predators is an important aspect of letting the wildness live in us. For those who haven't tasted the excitement of being the predator of a predator, it's exceedingly difficult to understand. No matter what you may think hunting is all about, the core value of hunting is certainly our union with the wild. A blog posting called "Go to the Wilderness" said:

> Those of us who live next to wild areas are living as everyone did when the world's major religions were established. Wilderness was right outside the door for most people in those days. Jesus, for instance, needed and loved the wilderness.
>
> Jesus was lucky, he had wilderness right outside his door. By the time of Jesus, people had already spread over the entire Earth, but wilderness was still accessible to nearly everyone.
>
> And Jesus needed it.[287]

Predators Can Be Fun

Coauthor Linda grew up in an area of Canada where bears and wolves were the main big predators. Wolves were shy, preferring the wilderness to anything human. But bears...well, bears like things that belong to humans, and they associate humans with food. Linda had to share the blueberry patch with bears and sometimes had to surrender a basketful to them. Keith, however, has a different idea about bears. Here's his story.

Playing with the Bears

Years ago, Linda and I hunted from a cottage-sized building that we call "The Cabin" located on our range property. (It's our class-room for the marksmanship courses that we do.) Off season, it's an excellent place to get away from the otherwise busy lifestyle our business demands. Each morning, during the hunting season, we would get up before daylight, enjoy one of Linda's delicious breakfasts, and launch for the day's hunt.

We were in the middle of the moose hunt, and on that morning, I decided to hunt on our east side. To get there I had to walk from the cabin to our six-hundred-meter range, walk up the range to the six-hundred-meter mound, and pick up a trail that would take me to the east side of our property.

Located between the five-hundred- and six-hundred-meter shooting mounds, we had a pile of carrots that we would be using as bait during deer season, and as I came onto the range, I could see a black bear sitting beside the pile with his back to me. He was relaxed and enjoying a meal of carrots. I noticed that the wind was blowing from me to the bear, and as I crossed the range at the three-hundred-meter mound, I wondered if he would get my scent.

The bear showed no signs of scenting me, so now I wondered just how close I could get to him before he did and what would he do then. I was walking down the tree line along the left edge of the range with the wind directly at my back. This meant my scent was not blowing directly at the bear but off to one side by about fifty meters. I thought this would be close enough for the bear to eventually smell my approach.

As I got to the five-hundred-meter mound, the bear still had not made any indication that he had winded me. I decided to cross to the other side of the range, causing the wind to blow my scent directly at the bear. I was just in front of the five-hundred-meter mound, so about sixty meters from the bear. As the bear munched on a carrot, he suddenly threw his head up, with his nose clearly searching the wind. He had been sitting on the ground with his back to me, but very quickly he was on his feet and running at top speed away from me, at a slight angle toward the tree line. He never looked back from the time he first scented me until he reached the tree line. It was obvious to him I was too close for a detailed look, and it was time to put more space between us.

I continued my hunt for moose, and by the end of the day, returning to The Cabin for the night, I was approaching the same carrot pile, this time from the other direction and with the wind in my face. I approached slowly and with caution, wondering if there would be a bear at the carrot pile. To my delight there was a sow and a single cub. The sow was lying in the middle of the carrots crunching on those carrots she could reach. The cub was playing nearby.

Now, with the wind in my favor, I wondered just how close I could get before they smelled me. There was a row of taller grass and weeds between us, so I got down on all fours and slowly made my way toward the bears. This row of weeds would put me about ten meters from them.

Once at the weeds, I watched the sow carefully, and when she looked away to see what her cub was doing, I stood up. I stood very still and was covered by the weeds from the waist down. The sow took a quick glance around but did not notice me and continued eating. Whenever she looked away, I took a sidestep along the weeds, putting me closer to her, and still, she did not notice me.

I was now within five meters of her, and I could feel the wind swirl. The sow was now sure there was something wrong. She froze in place, and while still lying down and without looking at me directly, she slowly turned her head away. She then suddenly exploded in the direction she was looking and took off running as hard as she could go, in fact, running so hard I could see pee flying everywhere.

The cub had been lying on his back, playing with his feet. The little guy was taken totally by surprise and was in complete wonderment as to why his mother took off so quickly without first gathering him up. He jumped to his feet and called after his mother, who was, by now, fifty meters away and still running hard. She didn't stop or answer his call, so he decided he needed to take off. His little legs didn't have a chance of catching up, and he ran as hard as he could to where he had last seen his mother disappear into the opposite tree line.

I continued to the cabin, wondering if I should tell Linda about my adventure with the bears. In the end, I did tell her and survived one of those looks I sometimes get.

When coauthor Dave read this story, he added:

> It is important to mention the possibility that the bear may have gone in the other direction, toward Keith, just as explosively. This is certainly not an activity advised for anyone without a high-power hunting rifle, lightning reflexes, great firearms skills, and maybe just a little death wish. And Linda's "look" was much deserved. But that is the point; that is the thrill.

At one level, this is what it means to be "wild" while still striving to be an ethical hunter. To know the thrill of being a predator, hunting a predator. For coauthor Dave, it was he and his dad in a swamp in Louisiana, up to their knees in water, with flashlights and .22 semi-auto rifles, in the dark, shooting venomous water moccasin snakes. These snakes were a kind of varmint, a deadly pest with no protection under any law, who were engaged in a weird "swarming" activity. After shooting countless snakes, Dave and his dad finally beat a hasty retreat and left with the joy of being truly alive and wild, just this one time.

When Keith read Dave's story, he said he'd rather take his chances with a thousand bears than climb into a swamp with even one water moccasin.

Many hunters who have spent much time in the wild have a story that goes something like these. The ones who survived will gladly share the tale.

It's Worth Fighting for Hunting

> In the introduction to *A Hunter's Heart*, Richard K. Nelson deftly summarizes the enduring value of hunting: Hunting provides the most important staple foods in our home, deepens my sense of connection to the surrounding natural world, and sharpens my awareness that I am an *animal*, not separate from my fellow creatures but twisted together with them in one great braidwork of life. Hunting brings me into the wild and brings the wild into me.[288]

In *Heartsblood*, David Petersen reflects on the sacred nature of our blood-bond with our fellow wild creatures:

> To hunt, kill, and devour the flesh of creatures wild and free is…the most natural possible exercise for body and spirit…honorable hunting is a spiritual sacrament, a humbling genuflection to our…design…as well as a sacred affirmation of our ancient blood-bond with the wildlings that for millions of years fed us [and] fed on us. Thus, were we created.[289]

And thus, we sustain our true nature through hunting.

Grounds for Happy Hunting: Who Needs Hunting?

Earlier, we talked about what the world might be like with and without hunting. We mentioned how hunting gives value to wild animals and the wildlands and how hunting funds and enables conservation and game management. Now, we explore the specific groups of people who might benefit from joining the hunt.

As Paul Shepard points out in *Thinking Animals*, we *share* the planet and its resources with animals:

Increasingly, what we do on the Earth will be done by design, and animals will not survive unless they are made to be part of that design. But why should we include them? We know that in *our* world nothing is free. In the future every animal on Earth will exist at some cost to humanity.[290]

Shepard says there are four main arguments for preserving animals: economics, ecology, ethics, and human growth and thought. He says the first three are inadequate: animals aren't as economically efficient as other choices; many animals aren't ecologically necessary; ethics hasn't improved people's behavior in other arenas. He concludes that we need to save them because they are what shape human growth and thought.

He ends with, "We will save them, if at all, because without them we are lost."[291]

People who do not have enough contact with the wild may not be able to understand Paul Shepard's point of view. They may already be too far gone. It's like arguing about what shape a ball is to a blind man. If he doesn't touch it, he can't see it.

While scrolling through the internet, coauthor Linda caught a short clip of a video. A mother was interviewing her three-year-old daughter:

Mother: "What's your favorite animal?"
Daughter: "Tiger!"
Mother: "What animal scares you the most?"
Daughter: "Tiger!"
Mother: "The tiger scares you the most and the tiger is your favorite animal?"
Daughter: "Yes," she said as she smiled and nodded.

Humans, Especially Antihunters, Need Hunting

The irony is that those who are in opposition to the hunt are the people who need it the most. It is because they are separate from the natural world that they despise wild things, including hunters. While they clamor about saving the wildlife, they in no way understand what *wild* life is. The hunter is one of the last guardians of the wild and the wild creatures who live in it.

> The successful hunt is a solemn event, and yet it is done in a spirit of joy. It puts modern man for a moment in vital rapport with a universe from which civilization tends to separate him in an illusion of superiority and independence.
>
> —Paul Shepard, as quoted in *Heartsblood*

Quoting the earlier mentioned 1980 study by Yale professor Dr. Stephen Kellert, David Petersen points out:

[Hunters] reported significantly greater interest in wilderness areas, in living in proximity to wildlife, and in backpacking, camping out, and fishing activities than nonhunters…Thus, despite the somewhat paradoxical fact that hunters killed wild animals… they were characterized by substantially greater affection, interest, and lack of fear of animals than nonhunters.[292]

The fact that hunters were characterized by substantially less fear of animals is interesting and important. If nonhunters fear animals, then they would likely be intimidated by, even fearful of, those of us who do not! It's therefore easy to see why nonhunters ease their fears by making animals "cute" and by giving them baby characteristics (like huge eyes and breathless voices). To those

332 → On Hunting

who admire animals (predator and prey alike) for what they really are, treating animals like babies is beyond disrespectful.

To better understand the realities of nature, antihunters need hunting.

Young People Need Hunting

> One of the first, and continuing, modern tragedies has been the replacement of the old man or woman as mentor and guide by written documents.
>
> —Paul Shepard, *The Tender Carnivore and the Sacred Game*

It takes a hunter to make a hunter.

There's a "new" wave beginning in psychology that suggests many young people have been made less resilient by being over-protected in their childhoods. Youngsters need to be challenged, to be encouraged to reach beyond their comfort zones, and to be given opportunities to be independent.

Most urban children don't have the opportunity to be responsible for something "real." Country children have chores that relate directly to the well-being of animals and the farm family. Hunters' children learn responsible handling of hunting arms as well as the ethics of respecting wild animals and the meat they provide.

Young people need hunting to prove their worth. When they make mistakes, they need time and patience to find a solution. When they can learn the lesson and move on, they develop strength and maturity.

In *Heartsblood*, David Petersen quotes the conclusion of a *Time* magazine story by Lance Morrow et al.: "Teachers and counselors report that kids who are taught to hunt responsibly are generally among the more mature and better-mannered—and saner—adolescents in the wilds of modern American culture."[293] Petersen's own conclusion is to the point:

Could it be…[that] these young hunters…[are adopting] a good and natural life?

Could it be that a wisely guided initiation to hunting helps to develop young people's ability to distinguish between a virtual world of culture made and a living world of nature born?

If so—and I believe in my heart this is true— then we need all the young hunters, and all the old mentors, we can get. They, all of them, are the past and the future of humanity.[294]

To ensure the future of humanity, young people need hunting.

What [youth] want, what they need, is meaningful personal challenge: rites of personal passage.

—David Petersen, *Heartsblood*

Women Need Hunting
More than any other identifiable group, women need hunting… and hunting needs women.

There have been many articles written about the rapid growth rate of a market segment called "female hunters." Most of the writers have no genuine understanding of why a woman would want to hunt. Most authors think that women have motivations different from men. We challenge that thinking. The women hunters we know fit the same profile as the men that have been endlessly surveyed. Women hunt because they want to feel the astonishingly clear union with nature (and with our ancient roots). They want to challenge themselves and become proven members of wild nature. And they want to take responsibility for delivering life-giving meat to the table.

Mary Zeiss Stange is one of the foremost proponents of women who hunt, and she expresses best how women can have a positive effect on the future of hunting:

> It is time to take an honest look at the world of nature and our human involvement in that world, including the blood that is on all our hands, in one way or another. Hunters may know how to do that better than anyone else. And women hunters are especially well poised to express the meaning of human involvement in nonhuman nature right now, in ways simultaneously quite new and very old.[295]

We can only hope she's right.

To provide depth and breadth to our understanding of nature, women need hunting.

Meet Erica

She's young. She's a woman. She used to be an antihunter.

In an article with the intriguing subtitle, "You Can Change Someone's Perception of Hunting," author Josh Honeycutt tells Erica's story. She came from an antihunting family and was also an antihunter until she met a boy who introduced her to "what the outdoors was truly about." Erica says that she always ate meat, but she just didn't fully realize where it came from. And she thought hunters just went out and killed. She didn't know they are part of herd management, the management that prevents overpopulation (and the resulting disease and car accidents overpopulation can cause).

It only took one deer hunt for Erica to find the hunter within. She says: "It's like I'm a whole different person. I can't describe it. I've gone from not being a morning person to loving getting up early. The satisfaction of supplying my own meat is incredible. It has become a way of life for me. It's all I think about nowadays."

Erica's advice in talking to antihunters is "Ask them to look at what they're eating. Open their eyes. Hunting isn't slaughtering animals. There is so much thought that goes into a hunt. Encourage them to go out with a hunter and experience the entire process. If they don't want to watch a hunt, have them check cameras, plant food plots, or try some other aspect of the hunt. Get them involved, if they are willing."

"I am very proud of who I am," she said. "I am proud of what I do."[296]

To ensure genuine pride in humanity's relation to nature, people like Erica need hunting.

Animals Need Hunting

> Wild animals don't need human "friends."
> Truly wild animals don't want human companionship.
> What they need and want is wildness.
>
> —David Petersen, *Heartsblood*

Most of us don't have contact with wild animals. But, as Paul Shepard points out in *Thinking Animals*, "That call of the wild lingers in the unconscious of people even in the most modernized societies, where it is commercialized as a media product in nature pictures, articles, or documentaries."[297]

The first toy most babies get is a stuffed animal.

Most people have had the experience of watching one of their children or grandchildren, dreamy-eyed, playing with plastic horses or being totally engaged with cartoon dogs. The people who design these toys may not understand why the kiddies relate to these animals, but they certainly know that kids do.

Keith and Linda tell the story of their granddaughters visiting during the deer hunt.

They were just little, maybe three and four years old. Their dad took them out to the backyard to see Grandpa's "meat deer" hanging from the buck pole. The next day, we took the family to our neighbor's deer farm to see the "petting deer." There, the girls enjoyed hand-feeding the deer. When we got home that night, the younger one wanted to see the "meat deer" again. She clearly knew there was a difference, like two different types of animal…and that's how we see it too.

The most likely explanation for why these children, and people in general, resonate with animals is because we're all "cut from the same cloth." From the anthropologist's view, humanity's formative years were when we were "teenagers" on the great scale of time. This is both humbling and insightful. As Dave Foreman wrote in *Keeping the Wild*, "I've come to suspect that self-centered humanists are actually incapable of imagining a time or a place without [modern] humans present."[298]

And of course, many such individuals can't grasp the idea that, for most times past, those humans were hunters. When our children play with their toy animals, they are playing with the drives, urges, and concepts that we've had from the dawn of time.

As Paul Shepard points out, "The remnant of the truly wild gets smaller and less wild."[299] The last tatters of the hunter-gatherer tribes and the deep-woods hunters understand this better than anyone else on this planet. There can be a big difference between the wild and the reserve park. Wild nature has better answers than anything man-made or "man-manipulated." Wild herds are better than game farms, and game farms are better than zoos. The wild nature within us knows these things.

There are lots of people who agree that we need to keep the *wild* in wildlife, and they are working toward making that happen.

These wild animals need us all, all of humanity, to work together to give them a place to be, well, *wild*.

To ensure effective wildlife management and conservation and all the economic benefits that result (especially for Indigenous communities), the world needs hunting.

To ensure their future is protected, all animals need hunting.

We Are a Precious Resource

I hunt…to reaffirm natural reality in a made world gone mad. I hunt to provide myself, my wife, and a few good friends with the gift of untainted wild flesh, won hard, with hands, head, and heart, from feral field or forest. I hunt to reconnect, physically and spiritually, to the timeless life-and-death drama that shaped the human body, the human mind, and the human needs for challenge, adventure, and passion.

—David Petersen, *Heartsblood*

We hunters are part of the secret to understanding the past, and we may well be the only way to navigate the future.

Nature is important, and we are the embodiment of wild nature. If you can understand the hunter, you can understand wild nature.

If you are the hunter, you are wild nature.

Come. Join the hunt.

We Are Clan

I have never been happier, more exhilarated, at peace, rested, inspired, and aware of the grandeur of the universe and the greatness of God that when I find myself in a natural setting not much changed from the way He made it.

—Former US President Jimmy Carter, "A Childhood Outdoors" in *A Hunter's Heart*

We have always felt an easy-going camaraderie with other hunters. While we may at times feel alienated from our neighbors, our elected officials, or other communities or interest groups, we always feel we can find common ground with another hunter. Rick Bass (environmental activist, award-winning author, and hunter) wrote "An Appeal to Hunters," in which he talked about other hunters...not just we humans...but all the other predators in the bush.

> I think many of us are feeling increasingly a certain
> cultural ostracism, a misunderstanding from a socie-
> ty that is frightened by our passion for hunting—and
> our passion for being in the woods. But the good
> thing about this alienation is that it makes the bond
> between those of us who do hunt that much stronger.
> We don't feel the need to explain or defend ourselves
> when in each other's company.[300]

We hunters are a clan. The clan crosses lines of religion, nationality, color, gender, and every other divisive category we might name. None of the categories matter. The only thing that matters is that we share a deep understanding of the hunt.

Some of our clan members see hunting as a religious experience, and we see it as a spiritual experience. It certainly is an experience that reaches to the depths of our roots and soars to the heights of our own heaven. We are joined together by our connection to the divine. As Jonathan Haidt wrote:

> I note that the greatest scholar of religion in psychol-
> ogy, William James, took a lone-believer perspective
> too. He defined religion as "the feelings, acts, and
> experiences of individual men in their solitude, so far

as they apprehend themselves to stand in relation to whatever they may consider the divine."[301]

In that sense, hunting is the religion of our clan.

We Are Human Heritage

Around the world there are many beautiful World Heritage Sites, places like the Great Barrier Reef in Australia, the Dinosaur Park in Canada, Sichuan Giant Panda Sanctuaries in China, Serengeti National Park in Tanzania, Yellowstone National Park in the USA, Victoria Falls in Zambia and Zimbabwe, and many, many more. To be designated a World Heritage Site, the candidate must meet one of ten criteria. We hunters could be a World Heritage "site" by meeting this one criterion: "to bear a unique or at least exceptional testimony to a cultural tradition or to a civilization which is living or which has disappeared."[302]

> I will remain what I am until I die, a hunter, and when there are no buffalo or other game I will send my children to hunt and live on prairie mice, for where an Indian is shut up in one place his body becomes weak.
>
> —Sitting Bull

We Are Keepers of the Skills

Our ancestors taught their youngsters to hunt. We now have a multi-generation gap, where not only are there billions of people on this planet who have never hunted, but most of them wouldn't know where to turn if they wanted to learn.

In Canada and most of the United States, we are blessed by being close to the wilderness. Still, many people are not inclined to go there. In fact, many urban people only leave the city to go to another city, traveling by taxis and airplanes. They have never been to the wilderness.

So, what's a child to do? Well, one of the things available is a credit course at the University of Manitoba (in central Canada), where students enroll in the Environmental Field Investigations course and learn how to hunt. Throughout the US and a few campuses in Canada as well, so far, Delta's University Hunting Program, launched in 2017, is designed to recruit hunters and educate future wildlife management professionals. Through it, Delta aims to teach wildlife management students who do not have a hunting background about the integral relationship between hunting and conservation by providing them with a hands-on waterfowl hunting experience. In addition to receiving classroom instruction, students must successfully pass a hunter safety course, attend a shooting skills instruction day, and participate in a hunt that culminates in a meal featuring the game harvested.[303]

Many states and provinces have special licensing programs for junior hunters. If you are a hunter, get involved in the junior hunter programs. Teach your children, grandchildren, nieces and nephews, and the children of your friends. If you're not a hunter but you'd like to be, do an internet search on "hunting education" and "hunting mentor" to find programs. Visit your local sporting goods store. Join a hunting group on social media and ask questions. Sign on for a guided hunt; it takes a little more money, but you'll have a great experience and learn a lot. Always remember, it is the money that you are willing to spend that makes it all economically viable.

Not all the billions of people on earth will be able to hunt. Realistically, not even a sizable minority will be able to hunt. This is what makes it work. A small number of hunters put a large value on being able to hunt. That money helps hunting to flourish, and hunting maintains the wilderness.

The "Namibia versus Kenya" trope is a concept that is, by now, familiar to the reader.[304] It is a hope for the future of hunting, the future of wilderness, and the future of our species and our civilization. Additionally, in most of North America, we are lucky enough to have hunting available and economically feasible just about everywhere. That is a very good thing. That is something we can cherish and strive to preserve. Whatever you end up paying, in the end you will count it money well spent.

The coauthors of this book have all brought new hunters on board. We have taught our children and are planning to teach our grandchildren. We have mentored young people, and we've helped new adult hunters get started. It's an immensely satisfying part of being a hunter: teaching, mentoring, and ultimately watching a newcomer become a genuine hunter.

Reverence

> What, then, is the path toward healing…if there is no return to the pre-Neolithic?… The path forward is not about sacrifice. It is about recovering what we have long ago sacrificed—our wholeness and our connection to other life and our deepest selves.
>
> —David Johns, "With Friends like These, Wilderness and Biodiversity Do Not Need Enemies" in *Keeping the Wild*

Most hunters have experienced a feeling of reverence, a profound connection to something bigger than themselves, a feeling of awe and respect, a sense of wonder and joy.

Sometimes we feel it just by being in the wilderness, wandering on ground that was here long before us and to which we will return. Sometimes we feel it while we're on the hunt: the focus of watching for sign, mindfulness of stalking our prey, knowing that there's a golden thread between us and the game we seek. Sometimes (often) we feel it at the moment of the kill…that heartbeat shared, that precious life given over to us, that deep appreciation. We have all felt the exhilaration mixed with sadness, that thanksgiving mixed with respect, and that bond mixed with awe. We are part of something big, and we celebrate the privilege of making contact.

These are our private moments, moments the hunter and the universe share. We show our reverence for wild nature in many other ways. Each year we renew our participation in game management. We support conservation and rewilding projects. We foster a deep understanding with the last fragments of hunter-gatherers before they are extinct. We share our knowledge and skills with new hunters, passing along our culture of reverence.

When others question us, we try to answer respectfully. We struggle with this at times. As David Petersen says, "Devout anti-hunters…crave a more literal, simplistic, predictable, personally compliant, and controllable world."[305] This is the opposite of reverence. But each of us tries, through our actions and our words, to demonstrate that hunting "speaks to me" and has value, relevance, and beauty. But for those who can't hear that message, we are saddened, but we accept them for who they are.

Celebrated hunter and conservationist Jim Shockey is well-known for providing guided hunts that respect the hunter and the hunted and for producing TV shows that reflect the same values. He acknowledges the spiritual side of hunting.

> Often we select our quarry based on the experience connected to it. We want to stay out longer, experience more in the wild outdoors…There is something spiritual about that, about sleeping under the stars, about trekking up mountains, about exerting yourself in nature.[306]

By experiencing the hunt, we experience both the history of humanity and the otherness of ourselves. We have touched the wildness in our genes, and it speaks to us. Whoever we become, wherever we may be, we will always crave "one more hunt."

> The otherness of stones and stars, like that of wild animals, is their deepest mystery, for they are not our products and their purposes are their own. They are the models for thinking our humbleness in the universe, and they are the key to the strangeness of ourselves.
>
> —Paul Shepard, *Thinking Animals*

Campfire Story: Her Daughter's First Deer

It was movement that caught her eye. She said nothing. She waited for the girl to see it on her own. The woman could see it was a magnificent buck, at least twelve points, tall and wide. It would be a grand deer for her daughter's first. She fondly remembered her own first deer, here in this same clearing.

Her grandfather had made her a rifle to fit her, and she had passed on that rifle to the girl who would now shoot her first deer with it. They were hunting from a hard stand that the old man had built to take the place of the pop-up hides. On one wall were pictures of all the deer that had been taken from this hut.

The buck lowered and raised its head, trying to get the scent of any danger. It then stepped forward, moving closer to the decoy. It was then that the girl saw the movement and recognized that this was her buck. She gasped softly as a shot of adrenaline hit her bloodstream. Her mother had told her this would happen, and she took several deep breaths to help settle down.

The girl forgot her mother was with her, as she focused on what she had practiced.

This deer had antlers, big antlers. *Breathe...breathe*, she told herself. The buck hadn't seen her and was far more interested in the decoy. *Breathe and do things right. Do things the way we rehearsed.*

Calmer now and in control, she reached for her ear defenders and put them on. Next, she picked up her shooting bipod and put it in position. She put the rifle in place.

The buck moved out into the clearing. It was magnificent and interested only in the decoy. Its ears were back, and it walked stiff legged. The girl now experienced a shot of adrenaline like never before. This was it! This was the buck! She had prepared herself for just this moment, and now she was shaking more than ever before. *Breathe. Breathe. Breathe!* she screamed in her mind. She could feel herself calming as she put her sight reticle on the point of aim.

She could see the sight reticle was steadier now, and as the big deer stopped to rock its head back and forth slightly, to make sure this decoy intruder could clearly see what it was about to tangle with, she was ready to fire her shot. Just like her mother had taught her: let the sight picture move around a little as long as it's within the kill zone. Apply a smooth steady squeeze to the trigger and wait for the rifle to go off.

She expected the sound of the shot to be louder, but that was the auditory exclusion she had read about. Even accounting for wearing ear defenders, the sound was much less than on the range. She immediately called her shot, taking a mental picture of exactly where the crosshairs were when the shot went off. It was a good shot—right on the point of aim. She reloaded the rifle.

The woman watched as the buck immediately did a mule kick and ran toward the nearest tree line, so closely echoing the actions of her own first deer decades ago in this very spot. If it got to the tree line, she knew her daughter would mark its entry by noticing which tree or bush it ran past, and she would be able to pick up the blood trail at that point. The deer slowed, becoming unsteady on its feet, and collapsed just as it neared the trees. The two could now just make out the one antler over the vegetation, and it soon stopped moving.

The girl now remembered her mother was with her, and she looked back for a reaction. Her mother smiled and nodded her head approvingly.

When it was time, the woman hugged the girl and told her to go check on her deer. The girl went alone, walking in the footsteps of her mother years before and performing the careful actions passed down to her. She kept her eyes on the antler she could see, her rifle in a "ready for anything" position as she approached. She circled slightly to approach the deer from its back and could smell the odor of a buck deer in full rut. She nudged it with her foot, and there were no signs of life. She then moved around to be able to look into its eyes and saw the stare of death.

She trembled slightly. She knelt down by its head and put her hand on its neck. She patted it gently and said, barely more than a whisper, "Thank you. Thank you for coming to me. Thank you for this experience. Thank you for the meat we'll have this winter. And thank you for these magnificent antlers. I will respect your death for the rest of my life." She sat quietly thinking about her first deer until she felt her mother move in beside her.

Buffalo Jump Revisited

An aboriginal guide was showing a group of tourists around [the exhibit at] Alberta's Head-Smashed-In Buffalo-Jump... The guide graphically described how in ancient times the buffalo would be driven over the edge of a fifteen meter precipice, to land in a gory heap at the base of the cliff. A diorama showed men and women clambering over the bodies to club and spear those still living.

When one tourist expressed shock at the bloody nature of the enterprise, the guide responded simply but with conviction, "We were hunters!" connecting her own generation with those of the past. She then amended her statement with equal conviction, adding, "Humans were hunters!" thus expanding complicity in the act of carnage to the whole of humanity [including the tourists].

—Richard B. Lee and Richard Daly,
The Cambridge Encyclopedia of Hunters and Gatherers

We were hunters. We *all* were hunters. And we are still.

If human history began twenty-four hours ago, less than six minutes ago, we were all still hunter-gatherers.

The Hunter's Prayer
by Linda K. Miller

I am a hunter hear my prayer.
It is not what I do; it is who I am.
I hold the power of life and death in my hand.
Let me act with purpose for a higher design;
   Keep me distant but not detached;
   Make me deliberate but not heartless.
For I am open to the way of the hunter,
And I am sworn to complete this journey.

I am your student; train my skills.
It is not what I do; it is who I am.
I have the power of determination in my heart:
Let me act with your guidance to use it well.
    Keep me focused on my task.
    Help me understand its final meaning.
For my mind is open to the way of the hunter.
And I am sworn to complete this quest.

I am your disciple; show me the path.
It is not what I do; it is who I am.
I have the power of bravery in my blood.
Let me act with courage and resolution.
    I am an agent of the greater good.
    Make my spirit pure and my will strong.
For my heart is open to the way of the hunter.
And I am sworn to complete this pursuit.

I am your agent; lead my way.
It is not what I do; it is who I am.
I have the power of lethal force in my means.
Let me act ethically and with integrity
    Make my choices clear and indisputable,
    Help me master my fate as I master my life.
For my soul is open to the way of the hunter,
And I am sworn to complete this passage.

I am a hunter; hear my prayer.
It is not what I do; it is who I am.
I hold the power of life and death in my hand.
And I am sworn to complete this journey.
Satis ver—Enough talk

Chapter Seventeen—Summary

Those who look from the outside can never really understand. The core value of hunting, at the heart of the hunter, is our *union* and *communion* with the wild. It is about being at one with the wilderness.

Most hunters have experienced a deep reverence...a profound connection to something greater than themselves. Hunters *know* the awe and respect, the wonder and joy. This transcendental experience is at its greatest when we participate directly in the cycle of life.

We feel it as we stalk our prey. Every sense *alive* and *straining* to feel, to hear, to see, to smell. We have a profound awareness of the golden thread in the fabric of life that connects us to our prey.

Often, we feel it at the moment of the kill. The heartbeat shared. The precious life given over to us—the deep awe mixed with profound appreciation—the astounding exhilaration mixed with sadness. Our thanksgiving blends with respect as we witness the mystery of death that gives us life.

And we understand that we, too, must someday make that transformation. We, too, will step across that line that separates life from death. Touching this ultimate mystery bestows in us a sense of peace in our own mortality.

We *all* have a faint sense of something vast. But the hunter celebrates the *privilege* of making contact with that immensity. These are our private moments. Moments shared by the hunter and the universe.

This experience is worth fighting for and worth sharing with others. And the beautiful wilderness in which we hunt is worth our time, money, energy, and effort to protect and restore.

The hunt sets us on a path to true kinship with our ancestors. We recognize and respect their lives and history as we fully utilize the skills of our heritage to become more holistic humans.

Hunting also makes us intimately in tune with death and therefore is of enormous value to every person as we approach the inevitable end of our own lifespan. Hunting is our best hope for preserving wilderness and habitat. Hunting is the path to a healthy future for our species and for our planet.

As roots are to a tree (vast, interwoven, nourishing, uplifting, and usually unseen), so is hunting to humanity. Hunting defines what it is to be human, and we cannot understand humanity without understanding hunting.

Acknowledgments

They say that writing is a lonely endeavor, and usually it is. But in the case of *On Hunting*, many people sustained us with their contributions through the eight years of making this book what it is today.

Many thanks to our editor, Rachel Libke, and her group of copyeditors. We never expected to have such a significant contribution made by any editor, but we do now understand and appreciate that Rachel is not "any editor." She's an outstanding professional.

Then there are those who contributed their thoughts and heart, whose words you'll read in the book. We are grateful for the endorsements we received and for the foreword written by Maggie Mordaunt. Maggie wrote straight from the heart and captured ours.

Three of our Campfire Stories were also written from the heart. Dave Siberry (a retired resort owner/operator) wrote "I Got an 8-Point Buck This Morning!," Viv Collings (at the time, a student) wrote "Viv's First Deer," and Sylvia Londry (a retired college professor) wrote "Sylvia's First Deer—A Textbook Shot!" Each of these hunters drew a memorable picture of their hunting experience so we could share it.

We also had a small group of alpha readers who waded through early drafts and moved the whole project forward. Aaron

Grubin not only provided some insight on aboriginal traditions of hunting in Canada but also included us in smudging rituals (first for the rifle that Keith built for him and later for his first deer, which he hunted here on our property). Amanda Scammell (college professor) provided a very careful edit and really helped us identify words, concepts, and sections that needed to be more user-friendly for nonhunters. Ian Crookston (engineer and competition shooter) read a close-to-final draft of the book and applied his thoroughness to this as he does to everything else in his life. He also shared some of what he witnessed in photo safaris, including the brutality of wildlife such as this: "It took over thirty minutes for the pack of hyenas to finally kill the baby buffalo. After the first minute, any human would be praying for a bullet—for themselves or any loved one. It was not a Disney moment."

We acknowledge the dear loved ones who first introduced each of us to hunting and the many authors whose books schooled us. We are particularly thankful for the authors who encouraged us and gave us free license to quote their work. And we thank the many people in our lives who emailed often to ask when they could buy a copy of *On Hunting*.

On a personal note…my biggest supporters were my coauthors. This was originally Dave's project. When he got us involved, he became our coach and champion and was very generous with his time and appreciation for our efforts—we are humbled and grateful. My partner and coauthor Keith wrote many of the stories and read many drafts but mostly waited as second in line behind the book for all the hours, days, and years that were spent on it.

To all of you I say this: I hope *On Hunting* is wildly successful. You have all contributed in ways that should make it so. From the bottom of my heart, I thank you.

—Linda K. Miller

Endnotes

1   Bruce Siddle, "Sharpening the Warrior's Edge," Keynote address, TREXPO East, August 2007.

2   Paraphrasing the value analysis pioneered and popularized by Dr. Morris Massey, "What you are is where you were when." "What You Are Is Where You Were When," *The Massey Triad* videotapes (Boston, MA: Enterprise Media, 1986).

3   "Persistence Hunting," *The Daily Omnivore* (website), August 2011, https://thedailyomnivore.net/2011/08/25/persistence-hunting/.

4   Robin McKie, "Humans Hunted for Meat 2 Million Years Ago," *The Guardian*, September 22, 2012, https://www.theguardian.com/science/2012/sep/23/human-hunting-evolution-2million-years.

5   *The Cambridge Encyclopedia of Hunters and Gatherers*, "Introduction: Foragers and Others," Richard B. Lee and Richard Daly (Cambridge University Press, 1999).

6   Robin McKie, "Did Stone Age Cavemen Talk to Each Other in Symbols?" *The Guardian*, March 10, 2012, https://www.theguardian.com/science/2012/mar/11/cave-painting-symbols-language-evolution.

7   McKie, "Did Stone Age Cavemen Talk to Each Other?"

8   David Petersen, "Hunting as Humanizer: Then and Now," *Center for Humans and Nature* (website), March 19, 2014, https://www.humansandnature.org/hunting-david-petersen.

9   Randall Haas, James Watson, et al., "Female hunters of the early Americas," *Science Advances* 6, no. 45 (Nov. 2020), https://www.science.org/doi/10.1126/sciadv.abd0310.

10   "Meeting Reports," Gordon Research Conference on Predator-Prey Interactions, January 2014, https://publish.uwo.ca/~lzanette/papers/zanette_et_al-2015-the_bulletin_of_the_ecological_society_of_america.pdf.

11   Andreas Löw, Peter J. Lang, J. Carson Smith, and Margaret M. Bradley, "Both Predator and Prey: Emotional Arousal in Threat and Reward," *Psychological Science*, 19.9 (September 1, 2008), 865–73.

12   Jeanna Bryner, "Modern Humans Retain Caveman's Survival Instincts," *Live Science*, September 24, 2007, https://www.livescience.com/4631-modern-humans-retain-caveman-survival-instincts.html.

13   Tom Marchant, "Time to Get Back to Your Roots?" *Forbes*, October 14, 2011, https://www.forbes.com/sites/juliahobbs/2011/10/14/time-to-get-back-to-your-roots/?sh=1c27ce8259e4.

14   Alanna Mitchell, "Why More Women Are Taking Up Hunting," *The Toronto Globe and Mail*, August 22, 2014, https://www.theglobeandmail.com/life/why-more-women-are-taking-up-hunting/article20179382/. Mitchell writes: "It is partly because of women that hunting is on the rise again after decades of decline. In 2011, according to a survey by the U.S. Fish and Wildlife Service, their numbers were up 36 per cent from 1991—versus a corresponding drop of 6 per cent in male hunters. If anything, the trend is even stronger in Canada. The number of women hunters in Alberta almost doubled between 2006 and last year [2013], while in Ontario it has risen 70 per cent in the past four years and in B.C. [British Columbia] by 62 per cent between 2003 and 2012. Last year, enrolment in Saskatchewan's mandatory hunter-education course rose by more than half over 2012, with women accounting for more than one-third of all students."

15   José Ortega y Gasset, *Meditations on Hunting* (Belgrade, MT: Wilderness Adventure Press, 2007), 105. Gasset (1883–1955) was a Spanish philosopher and author.

16   Sarah Klein, "Adrenaline, Cortisol, Norepinephrine: The Three Major Stress Hormones, Explained," *Huffpost*, April 19, 2013, www.huffpost.com/entry/adrenaline-cortisol-stress-hormones_n_3112800.

17   Wikipedia, s.v. "neurotransmitter," last modified September 11, 2021, https://en.wikipedia.org/wiki/Neurotransmitter. Dopamine plays a major role in reward-motivated behavior, as well as in motor control. Serotonin is believed to be a contributor to feelings of well-being, as well as regulating appetite and sleep and enhancing memory and learning.

18   Daniel Fisher, "Food on the Brain," *Forbes* (website), January 10, 2005, https://www.forbes.com/forbes/2005/0110/063.html?sh=4df58484de75.

19   Note that oxytocin is a naturally occurring hormone not to be confused with OxyContin (oxycodone), a narcotic pain reliever.

20   Dr. Sarah J. Buckley, "Pain in Labour: Your Hormones Are Your Helpers," *Natural Parent* Magazine, October 24, 2017, https://thenaturalparentmagazine.com/pain-labour-hormones-helpers/. Dr. Buckley writes: "Oxytocin helps us to be more emotionally open and more receptive to social contact and support. As the hormone of [female sexual pleasure], labour (it contributes to the euphoria of an unmedicated childbirth) and breastfeeding, oxytocin encourages us to 'forget ourselves,' either through altruism—service to others—or through feelings of love."

21   Jean-Claude Dreher, Simon Dunne, Agnieszka Pazderska, et al., "Testosterone Causes Both Prosocial and Antisocial Status-enhancing Behaviors in Human Males," *Proceedings of the National Academy of Sciences of the United States of America* 113, no. 41 (2016): 11633–38. doi:10.1073/pnas.1608085113.

22   Wikipedia, s.v. "Endorphins," last modified on October 10, 2021, https://en.wikipedia.org/wiki/Endorphins. From the website: "Endorphin is a contraction of endo- (body) and morphine, a morphine-like chemical produced by the body."

23   Peter Levine, "Dr. Peter Levine on Somatic Experiencing Approach and Titration," *PsychAlive* (website), https://www.psychalive.org/video-dr-peter-levine-somatic-experiencing-approach-titration/.

24   Anonymous user "Lineman," on CrossbowNation.com, May 10, 2013, https://www.crossbownation.com/threads/explanation-to-a-non-hunter-as-to-why-i-hunt.25069/.

25   Petersen, "Hunting as Humanizer: Then and Now."

26   Petersen, "Hunting as Humanizer: Then and Now."

27   Charles Patrick Davis, MD, PhD, editor, "Definition of Oxytocin," RxList (website), March 29, 2021, https://www.rxlist.com/oxytocin/definition.htm.

28   Jenny Morber, "What Science Says about Arousal during Rape," *Popular Science*, May 31, 2013, https://www.popsci.com/science/article/2013-05/science-arousal-during-rape/.

29   Jared Diamond, *Why Is Sex Fun?* (New York, NY: Perseus Books Group, 1997), 98.

30   Peter B. Gray, "Evolution and Human Sexuality," *Yearbook of Physical Anthropology*, 2013, https://www.academia.edu/5025235/Evolution_and_Human_Sexuality, 8.

31   See, for example, the 2014 Predator-Prey Interactions Gordon Research Conference in 2014 with the theme "From Genes to Ecosystems to Human Mental Health," https://www.grc.org/predator-prey-interactions-conference/2014/.

32   For example, see Eduard Machery, "Discovery and Confirmation in Evolutionary Psychology," in ed. J. Prinz *The Oxford Handbook of Philosophy of Psychology* (Oxford: Oxford University Press, 2011): "For example, although the importance of meat in hunter-gatherers' diet varies across societies, [studies] have shown that in all societies in their sample, the daily consumption of meat by hunter-gatherers is at least one order of magnitude larger than the daily consumption of meat by chimpanzees."

33   William R. Leonard and Marcia L. Robertson, "Evolutionary Perspectives on Human Nutrition: The Influence of Brain and Body Size on Diet and Metabolism," *American Journal of Human Biology*, volume 6.1, (1994) 77–78, https://onlinelibrary.wiley.com/doi/10.1002/ajhb.1310060111.

34   Ana Navarrete, Carel P. van Schaik, and Karin Isler, "Energetics and the Evolution of Human Brain Size," *Nature* (website), Springer Nature Limited/Nature Portfolio, November 11, 2011, https://www.nature.com/articles/nature10629. This research paper examines the "cognitive buffering hypothesis," which proposes that human brain capacity enabled effective responses to changing climates; additionally, meat sharing with mothers and children supported increased brain size.

35   Angela Wright, "Psychological Properties of Colours," Colour Affects (website), accessed January 1, 2019, http://www.colour-affects.co.uk/psychological-properties-of-colours.

36   Wright, "Psychological Properties of Colours."

37   Mike Parkinson, *Do-It-Yourself Billion Dollar Graphics* (Pepper Lip Press, 2006), 1.

38   Holle Kirchner and Simon J. Thorpe, "Ultra-rapid Object Detection with Saccadic Eye Movements," *Vision Research*, volume 46.11, (May 2006), 1762–76, at ScienceDirect, https://www.sciencedirect.com/science/article/pii/S0042698905005110.

39   Lt. Col. Dave Grossman with Loren W. Christensen, *On Combat: The Psychology and Physiology of Deadly Conflict*, (Milstadt, IL: Warrior Science Publications, 2008), 121.

40   "Persistence Hunting," *The Daily Omnivore*.

41   Daniel Lieberman et al., "Walking, Running and the Evolution of Short Toes in Humans," *Journal of Experimental Biology* 212, no. 5 (March 2009): 713–21. https://doi.org/10.1242/jeb.019885.

42   Maywa Montenegro, "The Running Man, Revisited," originally published in *The Seed Magazine*, March 18, 2009, found at South Willard (website), http://www.southwillard.com/news/the-running-man-revisited/.

43   "Oxytocin, Reciprocity, and Civil Society," *On the Commons* (website), October 24, 2005, http://www.onthecommons.org/oxytocin-reciprocity-and-civil-society#sthash.IXdp6mTY.PUXP65aw.dpbs.

44   J. Michael Plavcan and Carel P. van Schaik, "Interpreting Hominid Behavior on the Basis of Sexual Dimorphism," *Journal of Human Evolution* 32, no. 4 (1997): 371.

45   In *The Victim Cult: How the Culture of Blame Hurts Everyone and Wrecks Civilizations* (Victoria, BC: Thomas and Black, 2019), author Mark Milke eloquently describes those who identify as victims and the havoc they can raise individually and in society.

46   Rudyard Kipling (1865–1936), "The Law of the Jungle." You can read this poem in its entirety at http://www.kiplingsociety.co.uk/poems_ lawofjungle.htm.

47   For further reading, see cultural historian Christopher Lasch, *The Culture of Narcissism: American Life in an Age of Diminishing Expectations* (New York, NY: W. W. Norton & Company, 1979); Nathan Kowalsky, ed., *Hunting—Philosophy for Everyone: In Search of the Wild Life* (Chichester, West Sussex, UK: Blackwell Publishing Ltd. / John Wiley & Sons, 2010); David Petersen, *Heartsblood: Hunting, Spirituality, and Wildness in America* (Durango, CO: Raven's Eye Press, 2010); David Petersen, ed., *A Hunter's Heart: Honest Essays on Blood Sport* (New York, NY: Henry Holt and Company, 1997).

48   White-tailed deer don't distinguish any insects or grubs in their salads, and they've also been observed purposely eating birds and small mammals.

49   Hector Holthouse, *River of Gold: The Wild Days of the Palmer River Gold Rush* (Sydney, Australia: HarperCollins, 1967), 53.

50   Virginia Hughes, "When Do Kids Understand Death?" *National Geographic*, July 26, 2013, https://www.nationalgeographic.com/ science/article/when-do-kids-understand-death.

51   Jane Goodall, early in her research, noticed that chimps hunt and eat meat. Other scientific studies note that "although chimpanzees can and do hunt alone, they often form large hunting parties consisting of more than ten adult males, plus females and juveniles. Chimpanzees also go on 'hunting binges' in which they kill a large number of monkeys and other animals over a period of several days or weeks." Craig B. Stanford, "Chimpanzee Hunting Behavior and Human Evolution," American Scientist 83.3 (1995), 256–61.

Chimps are known to practice infanticide, and in June of 2012, an adult chimp killed a baby chimp at the Los Angeles Zoo. (https://latimesblogs.latimes.com/lanow/2012/06/baby-chimpanzee-killed-by-adult-chimp-at-la-zoo.html.) While this was shocking to zoo visitors, the behavior is not uncommon among male chimps who are interested in breeding a mother chimp.

These aggressive chimpanzee behaviors are likely driven from a need to feed and breed, sustaining the individual and his gene pool.

52 Most countries have strict rules about the hunting of these animals, and while fund-raising for game management is definitely a major part of licenses, the main objective is to ensure that the species endures.

53 In the first chapter of *Sapiens: A Brief History of Humankind* (New York, NY: Harper, 2015), Yuval Noah Harari provides an enjoyable description of how our ancestors may have developed and migrated, covering the time period from the dawn of humankind up to the Agricultural Revolution.

54 Esther Fleming, *SidmartinBio* (website), October 4, 2019, https://www.sidmartinbio.org/what-is-the-population-of-all-the-animals/.

55 The total surface of the earth is 197.4 million square miles. The land area is 57.5 million square miles or 29 percent of the total surface. The habitable land is estimated to be twelve million square miles or 3.5 percent of the total surface.

56 Kathy Andrews and Tracy Shafer, "More Than Bragging Rights," *Outdoor Illinois*, Illinois Department of Natural Resources, March 2006, https://www2.illinois.gov/dnr/OI/Documents/March06BigBuck.pdf. You can also find information about the Boone and Crockett Club and its mission of conservation at https://www.boone-crockett.org/.

57 "Rhino Hunter Culls Problem Bull in Namibia," *The Outdoor Wire* (website), Dallas Safari Club, May 26, 2015, https://www.theoutdoorwire.com/story/1432256678v4yu354br29.

58 Jacob Wawatie and Stephanie Pyne, "Tracking in Pursuit of Knowledge: Teachings of an Algonquin Anishinabe Bush Hunter," in *Hunting—Philosophy for Everyone*, 94.

59   Gary Snyder, *The Practice of the Wild* (Berkley, California: Counterpoint Press, 1990), 93–94.

60   For example, see Hugh Brody, *The Other Side of Eden: Hunters, Farmers, and the Shaping of the World* (New York, NY: North Point Press, 2001).

61   Roger Scruton, "The Sacred Pursuit: Reflections on the Literature of Hunting," in *Hunting: Philosophy for Everyone*, 185–97.

62   James Carmine, "Off the Grid: Rights, Religion, and the Rise of the Eco-Gentry," in *Hunting—Philosophy for Everyone*, 241.

63   Ben Moyer, "Hunting: Number of Hunters Is Dropping, but Not Public Support for Those Who Hunt," *Pittsburgh Gazette*, June 30, 2007, https://www.post-gazette.com/sports/hunting-fishing/2007/06/30/Hunting-Number-of-hunters-is-dropping-but-not-public-support-for-those-who-hunt/stories/200706300154.

64   Responsive Management & National Sports Shooting Foundation, "Americans' Attitudes toward Hunting, Fishing, Sport Shooting, and Trapping," 2019, https://www.fishwildlife.org/application/files/7715/5733/7920/NSSF_2019_Attitudes_Survey_Report.pdf.

65   Leading the way have been our First Nations. See Kelly Geraldine Malone, "'Go Back to the Old Way:' First Nations Return to Land during COVID-19 Pandemic," *National Post*, May 10, 2020, https://nationalpost.com/pmn/news-pmn/canada-news-pmn/go-back-to-the-old-way-first-nations-return-to-land-during-covid-19-pandemic. As another example, see the following report from Wisconsin citing increased deer tag sales and hunting supplies scarcity: "More people hunting during the pandemic," NBC26 Green Bay, September 23, 2020, https://www.nbc26.com/news/local-news/more-people-hunting-during-the-pandemic.

66   This is called a "predatory response," and a good example is the black bear who will chase you if you run. The Algonquin Park Bear Safety Rules state that if you encounter a black bear: "Do not turn and run—this may trigger a predatory response in the bear." See http://www.algonquinpark.on.ca/visit/recreational_activites/black-bear-safety-rules.php.

67   Ivan Carter, "A subject of much recent debate—
ban hunting and say goodbye," Facebook, January 7,
2016, https://www.facebook.com/173228692766778/
photos/a.232889763467337/944866072269699.

68   "Pleasure and Leisure in Ancient Persia," Gateways to the Ancient
Near East, May 11, 2012, https://ancientpersiansociety.wordpress.
com/2012/05/11/hello-world/.

69   Daisy Dunn, "The Elites of Ancient Rome Transformed the Nature of
Hunting," *History Today* 64, no. 2 (2014), historytoday.com/archive/
roman-hunt#sthash.1qBbwjfB.dpuf.

70   Edward H. Schafer, "Hunting Parks and Animal Enclosures
in Ancient China," *Journal of the Economic and Social History
of the Orient* 11, no. 3 (October 1968): 318–43, http://doi.
org/10.2307/3596278.

71   Simon Shadow, "How Poaching Works," *HowStuffWorks* (website),
InfoSpace Holdings, LLC, December 9, 2008, https://adventure.
howstuffworks.com/outdoor-activities/hunting/traditional-methods/
poaching1.htm.

72   According to reports from wildlife organization Save the Elephants,
the price (in 2014) for raw ivory in China was $2,100 per kilogram.
Back in 2010, the price of the ivory was $750 per kilo, but it has
dropped to $730 as recently as 2017 since China has pledged to ban
ivory trade. See Julia Zorthian, "The Price of Ivory in China Plunged
65 percent in the Last Three Years," *Time* Magazine, March 30, 2017,
https://time.com/4718327/china-ivory-prices-dropped/.

      In Cambodia, one of the most popular wildlife that is trafficked
across the world is the pangolin. According to wildlife conservation
officials, the meat of a pangolin is sold on the black market for $350
per kilogram. The pangolin's scales, which are used for medicinal
purposes, are sold for $600 to $1,000 per kilogram. See John Sutter,
"The Most Trafficked Mammal You've Never Heard Of," CNN,
April 2014, https://www.cnn.com/interactive/2014/04/opinion/sutter-
change-the-list-pangolin-trafficking/index.html.

73 "Bushmeat," International Affairs, U.S. Fish and Wildlife Service, accessed October 2, 2021, https://www.fws.gov/international/wildlife-without-borders/global-program/bushmeat.html.

74 "Fair Chase Statement," The Boone and Crockett Club (website), accessed October 4, 2021, https://www.boone-crockett.org/fair-chase-statement.

75 Theodore R. Vitali, "But They Can't Shoot Back: What Makes a Chase Fair," in *Hunting—Philosophy for Everyone*, 28. Vitali, CP, PhD, is an Associate Professor and former Chair of Philosophy at Saint Louis University. He is also a professional member of the Boone and Crockett Club.

76 Aldo Leopold (1887–1948), who spent his childhood appreciating the great outdoors, put his interests to work in the fields of science, ecology, forestry, conservation, and environmentalism. He was also an author, a professor of agriculture, and a research director at the University of Wisconsin.

77 Jonathan Haidt, *The Righteous Mind: Why Good People Are Divided by Politics and Religion* (New York, New York: Pantheon Books, 2012), 336.

78 Henry David Thoreau, *Walden, or Life in the Woods* (New York, NY: Signet Classics, 2004), 169. Thoreau (1817–1862) was an influential writer, naturalist, and historian whose philosophies inspired many conservationists and activists.

79 Theodore Roosevelt, *Outdoor Pastimes of an American Hunter* (New York, NY: Charles Scribner's, 1905), 272.

80 Petersen, *Heartsblood*, 34–35.

81 Aldo Leopold, *A Sand County Almanac and Sketches Here and There* (New York, New York: Oxford University Press, 1968), 204.

82 Some people would describe this as a "flow experience," as articulated by Mihaly Csikszentmihalyi. See Mike Oppland, "8 Ways to Create Flow according to Mihaly Csikszentmihalyi," Positive Psychology, August 12, 2021, https://positivepsychology.com/mihaly-csikszentmihalyi-father-of-flow/.

83   Glen Martin, "Kenya: Wildlife Loved to Death," *African Indaba*, Volume 12.2 (March 2014), International Council for Game and Wildlife Conservation, http://www.africanindaba.com/2014/03/kenya-wildlife-loved-to-death-march-2014-volume-12-2/.

84   Glen Martin, "Lionizing Cecil Makes Us Feel Good, but a Trophy Hunting Ban Will Accelerate Slaughter," *California* (website), CalAlumni Association at U.C. Berkeley, August 3, 2015, https://alumni.berkeley.edu/california-magazine/online/lionizing-cecil-makes-us-feel-good-trophy-hunting-ban-will/.

85   Brody, *The Other Side of Eden*, 247.

86   "Theoretical Tiger Chases Statistical Sheep to Probe Immune System Behavior: Physicists Update Predator-Prey Model for More Clues on How Bacteria Evade Attack from Killer Cells," *ScienceDaily*, April 29, 2016, https://www.sciencedaily.com/releases/2016/04/160429095152.htm.

87   "The brain capacity [of these ancestors] was about 1,600 cc (98 cu in), which is *larger* than the average for modern humans." *Walking with Wikis* (website), s.v. "Cro-Magnon," accessed October 4, 2021, https://walkingwith.fandom.com/wiki/Cro-Magnon.

88   Valerius Geist, "The Carnivorous Herbivore: Hunting and Culture in Human Evolution," in *Hunting—Philosophy for Everyone*, 128.

89   Marlene Cimons, "'Social' Bacteria That Work Together to Hunt for Food and Survive under Harsh Conditions," National Science Foundation (website), December 20, 2013, https://www.nsf.gov/discoveries/disc_summ.jsp?cntn_id=129976.

90   "Poisons Used by African Tribes for Hunting," Gateway Africa, accessed October 4, 2021, https://www.gateway-africa.com/howdidthey/poisonforarrows.html.

91   Heather Pringle, "Ice Age Communities May Be Earliest Known Net Hunters," *Science* 277 (August 1997): 1203–4, https://www.academia.edu/2443569/ARCHAEOLOGY_Ice_Age_Communities_May_Be_Earliest_Known_Net_Hunters.

92   "Alligator Snapping Turtle," National Geographic (website), accessed
     October 4, 2021, https://www.nationalgeographic.com/animals/
     reptiles/facts/alligator-snapping-turtle.

93   Mary Zeiss Stange, *Woman the Hunter* (Boston, Massachusetts:
     Beacon Press, 1997), 187.

94   "Komodo dragon," National Geographic (website), accessed October
     5, 2021, https://www.nationalgeographic.com/animals/reptiles/facts/
     komodo-dragon.

95   Brian Resnick, "The Animal Kingdom's Top Marathoners,"
     Popular Mechanics, Nov. 5, 2010, https://www.popularmechanics.
     com/adventure/sports/g418/animal-kingdom-top-marathon-
     runners/?slide=3. Eliud Kipchoge (Kenya) holds the current (as of
     January 2022) world record at 2:01:39 (Berlin, 2018) as reported in
     runnersworld.com.

96   Paul Shepard, *The Tender Carnivore and the Sacred Game* (Athens,
     Georgia: University of Georgia Press, 1998), 114.

97   Shepard, *The Tender Carnivore and the Sacred Game,* 154.

98   Shepard, *The Tender Carnivore and the Sacred Game,* 155.

99   Shepard, *The Tender Carnivore and the Sacred Game,* 156.

100 Shepard, *The Tender Carnivore and the Sacred Game,* 159.

101 "Zoo Chimp 'Planned' Stone Attacks," *BBC News,* March 9, 2009,
     http://news.bbc.co.uk/1/hi/sci/tech/7928996.stm.

102 Victoria Gill, "Oldest Evidence of Arrows Found," *BBC News,* August
     2010, https://www.bbc.com/news/science-environment-11086110.

103 This behavior may be why your pet cat brings you a somewhat
     disabled mouse or mole. Perhaps it's not so much a gift as it is a
     teaching aid. Is your cat trying to tell you that you really do need to
     grow up and start doing your own hunting?

104 For an interesting list, see *List of Fatal Bear Attacks in North America*
     posted at https://en.wikipedia.org/wiki/List_of_fatal_bear_attacks_
     in_North_America#2020s.

105   Bryan Nelson, "15 Cute Animals That Could Kill You," TreeHugger, July 21, 2020, https://www.treehugger.com/cute-animals-that-could-kill-you-4869354.

106   Ronald M. Nowak, *Walker's Mammals of the World* (Baltimore, MD: The Johns Hopkins University Press, 1999), 792.

107   The Tsavo man-eaters publicized in the 1996 movie *The Ghost and the Darkness* were prolific man-hunters. There are many examples of man-eating lions, and modern research now acknowledges that this behavior is not aberrant but quite normal for the big cats. There may be specific conditions, such as a lack of usual prey or an abundance of human corpses, that trigger the behavior. But once these predators find a source of easy prey, it only makes sense for them to pursue it.

108   Peter Hathaway Capstick, *Death in the Silent Places* (New York, NY: St. Martin's Press, 1981), 189.

109   Siddle, "Sharpening the Warrior's Edge."

110   Rick Bass, "An Appeal to Hunters," in *A Hunter's Heart*, 194.

111   TyB, "10 Truly Awful Ways to Be Killed by an Animal," *ListVerse*, July 5, 2010 https://listverse.com/2010/07/05/10-truly-awful-ways-to-be-killed-by-an-animal/.

112   Two fatal coyote attacks have been confirmed by experts: On August 26, 1981, a coyote attacked three-year-old Kelly Keen in the driveway of her Glendale, California, home and ran off with her. She was rescued by her father and rushed to Glendale Adventist Medical Center but died in surgery due to a broken neck and blood loss. On October 27, 2009, Taylor Mitchell, a nineteen-year-old Canadian folk singer on break from a concert tour, died from injuries and loss of blood suffered in an attack by two coywolves at Cape Breton Highlands National Park's Skyline Trail in Nova Scotia. See Wikipedia, s.v. "Coyote Attack," last modified September 19, 2021, https://en.wikipedia.org/wiki/Coyote_attack.

113   A concern of biologists in North America these days is that the modern coywolves (bigger than a coyote, more social than a wolf) will become more of a threat as they infiltrate the suburban areas. See Bob Humphrey, "Coydog, Coywolf, or Coyote? The 5 Things You Need to

Know about Eastern Canids," *Outdoor Life*, March 9, 2021, https://www.outdoorlife.com/story/hunting/eastern-coyote-facts/.

114 Valerius Geist, "Beyond Wolf Advocacy, toward Realistic Policies for Carnivore Conservation," Boon and Crockett Club (website), Winter 2008, https://www.boone-crockett.org/beyond-wolf-advocacy-toward-realistic-policies-carnivore-conservation.

115 "Never Cry Wolf," International Wolf Center, accessed May 19, 2022, https://wolf.org/original-articles/never-cry-wolf/.

116 Wikipedia, s.v. "Wolf Attack," last modified October 4, 2021, https://en.wikipedia.org/wiki/Wolf_attack.

117 Mark E. McNay, "A Case History of Wolf-Human Encounters in Alaska and Canada," 2002, i. https://digitalcommons.unl.edu/cgi/viewcontent.cgi?article=1025&context=wolfrecovery.

118 For a list that includes this and other wolf attacks, see Wikipedia, s.v. "List of Wolf Attacks in North America," last modified July 19, 2021, https://en.wikipedia.org/wiki/List_of_wolf_attacks_in_North_America.

119 "Hunting," The Jane Goodall Institute UK, accessed October 16, 2021, https://www.janegoodall.org.uk/chimpanzees/chimpanzee-central/15-chimpanzees/chimpanzee-central/20-hunting.

120 Craig B. Stanford, "The Predatory Behavior and Ecology of Wild Chimpanzees," 2011, archived at *Gambassa* (website), accessed October 6, 2021, http://www.gambassa.com/public/Collaborations/1264/850/4911/Stanford,2011.html.

121 Stanford, "The Predatory Behavior and Ecology of Wild Chimpanzees." Dr. Stanford's article starts with this: "When Jane Goodall first observed wild chimpanzees hunting and eating meat nearly 40 years ago, skeptics suggested that their behavior was aberrant and that the amount of meat eaten was trivial. Today, we know that chimpanzees everywhere eat mainly fruit, but are also predators in their forest ecosystems. In some sites the quantity of meat eaten by a chimpanzee community may approach one ton annually. Recently revealed aspects of predation by chimpanzees, such as its frequency and the use of meat as a political and reproductive tool, have important implications for research on the origins of human

behavior. These findings come at a time when many anthropologists argue for scavenging rather than hunting as a way of life for early human ancestors. Research into the hunting ecology of wild chimpanzees may therefore shed new light on the current debate about the origins of human behavior." Researchers have also observed that bonobos eat meat. See Monica L. Wakefield, Alexana J. Hickmott, et al., "New Observations of Meat Eating and Sharing in Wild Bonobos (*Pan paniscus*) at Iyema, Lomako Forest Reserve," *National Library of Medicine*, National Center for Biotechnology Information, March 19, 2019, https://pubmed.ncbi.nlm.nih.gov/30889597/.

122 Brian Palmer, "Do Wolves Kill for Sport?" *Slate*, November 05, 2009, https://slate.com/news-and-politics/2009/11/do-wolves-kill-for-sport.html.

123 Palmer, "Do Wolves Kill for Sport?"

124 Benjamin Taub, "Why Are Herbivorous Prairie Dogs So Murderous?" *IFL Science* (website), March 24, 2016, https://www.iflscience.com/plants-and-animals/why-do-some-herbivores-kill-other-animals/.

125 Dennis Stapleton, "Moose vs. Bear: Who Wins in a Fight?" *Misfit Animals*, accessed May 19, 2022, https://misfitanimals.com/bees/moose-vs-bear/.

126 Byakko Toranosuke, "Top 10 Herbivores You Probably Want to Avoid," *ListVerse*, January 10, 2010, https://listverse.com/2010/01/10/top-10-herbivores-you-probably-want-to-avoid/.

127 Toranosuke, "Top 10 Herbivores You Probably Want to Avoid."

128 Shepard, *The Tender Carnivore and the Sacred Game*, 196.

129 Shepard, *The Tender Carnivore and the Sacred Game*, 198.

130 Baiting is legal in this jurisdiction, and carrots were recommended by the local game warden.

131 "Seeing Movement: Why the World in Our Head Stays Still When We Move Our Eyes," Max–Planck–Gesellschaft, The Max Planck Society, March 21, 2012, https://www.mpg.de/5038240/brain_seeing_movement.

132 Sometimes the caliber is determined by the game laws of the area or the country where you are planning to hunt. For example, African countries require a minimum caliber for dangerous game.

133 Jim Carmichel, "Knockdown Power: Here's Why Some Calibers Always Seem to Flatten Game," *Outdoor Life* (website), August 2003, https://www.outdoorlife.com/articles/hunting/2015/04/knockdown-power-heres-why-some-calibers-always-seem-flatten-game/.

134 Combat breathing routine: breathe in deeply for a count of four, hold for a count of four, breath out for a count of four, hold for a count of four. Repeat until you're feeling calm.

135 The mental aspects of marksmanship are so important that we (Linda K. Miller and Keith A. Cunningham) wrote a whole book about it. *Secrets of Mental Marksmanship* (Mascoutah, IL: Killology Research Group, 2018) discusses many mental skills, each from the point of view of the hunter, the military, the law enforcement officer, and the competitor.

136 Miller and Cunningham, *The Wind Book for Rifle Shooters* (Manhattan, NY: Simon and Schuster, 2020) is a popular book that covers wind-reading, from the basics to the techniques used by expert competitors around the world.

137 Internet-based map products like Google Earth are also useful. However, we usually use paper because of connection problems in our somewhat remote location. Drones and trail cameras can be helpful, but be aware of local regulations regarding their use.

138 Alison Acton, "Getting By with a Little Help from My Hunter," in *Hunting—Philosophy for Everyone*, 84.

139 Varmint-hunting often goes by different rules than game hunting.

140 Carlos Gallinger, "Hunting in the Human Way: The Shaman and Hunting Magic," The Way of Things (website), January 26, 2013, https://thewayofthings.org/philosophy/hunting-human-way-shaman-and-hunting-magic.

141 Gallinger, "Hunting in the Human Way."

142 Grossman, *On Combat*, 43–44.

143 This story by Linda Miller and Keith Cunningham was originally published in *Ontario Out of Doors* magazine, Jan–Feb 2016.

144 An old blue is an eland that is very old and consequently has a blueish tinge to his neck and shoulders.

145 In *The Victim Cult*, Mark Milke makes the case that some of the most egregious acts of war were popularized by leaders who positioned themselves and their countrymen as victims, and so their war was in defense of their lives and their way of life.

146 Carl von Clausewitz, *On War vol. III* (London: N. Trubner and Co., 1873,), 85–86.

147 "Criminal Justice, Section 4.1: Early History," *Lumen* (website), accessed April 1, 2022, https://courses.lumenlearning.com/atd-bmcc-criminaljustice/chapter/section-4-1-early-history-of-policing/.

148 Marilee Hanson, "Roman Soldiers—History and Facts," February 14, 2022, https://englishhistory.net/romans/roman-soldiers/.

149 Elizabeth Nix, "Why Are British Police Officers Called 'Bobbies'?" History.com, August 22, 2018, https://www.history.com/news/why-are-british-police-officers-called-bobbies.

150 Barbara Ehrenreich, *Blood Rites: Origins and History of the Passions of War* (New York, NY: Grand Central Publishing, 2020), 135. The author is quoting Ruth Benedict in *The Natural History of War*.

151 Ehrenreich, *Blood Rites*, 83–84.

152 Jared Diamond, *The Third Chimpanzee: The Evolution and Future of the Human Animal* (New York, NY: HarperCollins, 1992), 296.

153 Diamond, *The Third Chimpanzee*, 308.

154 Diamond, *The Third Chimpanzee*, 308.

155 The Qur'an similarly prohibits wanton killing. "And do not take any human being's life—that God willed to be sacred—other than in [the pursuit of] justice" (Q 17:33).

156 "Thou shalt not kill" comes from the King James translation of the Bible. Let's cut right to the chase: "Thou shalt not kill" in Exodus 20:13 is a flat-out, incorrect translation. The context of the rest of that chapter makes it very clear. And in Matthew 19:18 in the King James

translation, Jesus is reciting the Commandments, and he says, "Thou shalt do no murder."

157 *The Third Chimpanzee* mentions massacres intended to capture a neighbor's territory, food, or females are conducted by lions, wolves, hyenas, ants, langur monkeys, gorillas, and chimpanzees. Diamond, *The Third Chimpanzee*, 290.

158 Diamond, *The Third Chimpanzee*, 296–97.

159 Grossman, *On Killing*, 4.

160 Milke, *Victim Cult*, 120–50.

161 Diamond, *The Third Chimpanzee*, 300.

162 *On Killing* has been often cited in scholarly works (over 3,200 times as of 2022, according to Google Scholar).

163 Grossman, *On Killing*, 29, 254.

164 Gwynne Dyer, *War: The New Edition* (Toronto, ON: Random House of Canada, 2010), 57.

165 Grossman, *On Killing*, 189.

166 Grossman, *On Killing*, 179.

167 Keith was in the US Army during the Vietnam War then stayed on in the USA for GI-Bill training (as a gunsmith). He worked in the US for a few years before returning to Canada, where he joined the Canadian Army.

168 Dyer, *War*, 53.

169 Grossman, *On Killing*, 180–81.

170 For more information on post-traumatic stress (PTS) and post-traumatic stress disorder (PTSD), which are *not* the same thing, the reader is encouraged to read Lt. Col. Dave Grossman's book *On Combat*, which is used for understanding these psychological dynamics in many different scholarly, medical, and practical environments.

171 This color code system was popularized by Lieutenant Colonel Jeff Cooper, *Principles of Personal Defense* (Boulder, Colorado: Paladin Press, 2006).

172 For a clinical look at the mechanics of brain function and sensory distortions, see *The Emotional Brain* by Joseph LeDoux. He discusses how the amygdala can trigger an emotional reaction to something well before the cerebral cortex has had a chance to process the event. This may be an aspect of "déjà vu," and it may be a contributor to sensory distortions in high-stress situations.

173 Grossman, *On Combat*, 120.

174 Haidt, *The Righteous Mind*, 206.

175 In *The Righteous Mind*, Haidt gives an excellent example from hen breeding and egg production. He says that if you breed only the most productive hens (individual selection), the total egg production drops because of the hens' aggressive behavior, which includes killing and cannibalism. But if you use group selection, breeding the whole cage of twelve hens that were in the most productive cages, then aggression levels drop and total egg production increases…and every hen is more productive.

176 Ehrenreich, *Blood Rites*, 82.

177 Haidt, *The Righteous Mind*, 221.

178 Haidt, *The Righteous Mind*, 365.

179 Ehrenreich, *Blood Rites*, 117–18.

180 Grossman, *On Combat*, 212.

181 Some of this was what the military refers to as "recon by fire," and some was massive amounts of fire such as Gatling guns shooting from aircraft, firing thousands of rounds per minute, but there can be no doubt that the overall hit ratio was truly embarrassing and abysmal.

182 Lt. Col. Dave Grossman's book *On Combat* has been cited over six hundred times in scholarly works (according to Google Scholar as of 2022).

183 Grossman, *On Combat*, 170.

184 Grossman, *On Combat*, 170.

185 Grossman, *On Combat*, 170.

186 Miller and Cunningham, *Secrets of Mental Marksmanship*.

187 Jeff Cooper, *To Ride, Shoot Straight, and Speak the Truth* (Summerville, MA: Wisdom Publications, 2005), 20.

188 Cooper, *Principles of Personal Defense*, 42.

189 Cooper, *Principles of Personal Defense*, 60–61.

190 Joseph Boyden, *Three Day Road* (Toronto, ON: Penguin Random House Canada, 2005). The story was inspired somewhat by real-life Ojibwa hero Francis Pegahmagabow.

191 Patrick van Horne and Jason A. Riley, *Left of Bang* (New York: Black Irish Entertainment, LLC, 2014), 18.

192 Although the program is designed to enhance the soldier's predator skills, the objective is not to annihilate but to control…so, a successful outcome doesn't necessarily require killing. In that sense, the predatory behavior is much like the animal who is controlling its territory or its harem.

193 Van Horne, *Left of Bang*, 22.

194 Van Horne, *Left of Bang*, 31. Van Horne is quoting from David G. Myers, "Intuition: Its Powers and Perils." Psychological Inquiry 21 (2010): 371–77, https://www.jstor.org/stable/25767208.

195 Chris Bird, *Thank God I Had a Gun: True Accounts of Self-Defense* (San Antonio, Texas: Privateer Publications, 2007), 1–38.

196 Haidt, *The Righteous Mind*, 223.

197 Grossman, *On Combat*, 51.

198 This information is from an interview with a photo safari guide working with Bush and Beyond, a safari company in Kenya.

199 Nelson, "Finding Common Ground," 3.

200 Stephen Kellert, "Attitudes and Characteristics of Hunters and Antihunters," *International Journal for the Study of Animal Problems* (1980): 87–119.

201 Brody, *The Other Side of Eden*, 294.

202 Brody, *The Other Side of Eden*, 294–300.

203 Thomas McIntyre, "What the Hunter Knows," in *A Hunter's Heart*, 177–81.

204 "The term *carrying capacity* is applied largely to animal populations as the maximum number of individuals in a particular species that can be indefinitely supported by the resources in a particular area. In animal contexts, the carrying capacity is determined by the amount of food available, the number of predators, and the rate at which the environment can replace the resources that are used by the population. In applying this concept to humans there are two differences. First, human beings have the capacity to innovate and to use technology and to pass innovations on to future generations, so they have the capacity to redefine upward the limits imposed by carrying capacity. Second, human beings need and use a wider range of resources than food and water in the environment. Hence, human carrying capacity is a function of the resources in an area, the consumption level of those resources, and the technology used in exploiting them." Graeme Hugo, Professor of Geography, University of Adelaide, Australia, "Projections of Global Carrying Capacity," *The Role of Food, Agriculture, Forestry and Fisheries in Human Nutrition* (Oxford, UK: EOLSS Publishers/UNESCO, 2011), 31, https://www.eolss.net/Sample-Chapters/C10/E5-01A-04-01.pdf#.

205 Shepard, *The Tender Carnivore and the Sacred Game*, 240–41.

206 Kay Koppedrayer, "Big Game and Little Sticks: Bowmaking and Bowhunting," in *Hunting—Philosophy for Everyone*, 204.

207 Stange, *Woman the Hunter*, 121.

208 Steve Sanetti, "Hunter Green," *The Washington Post*, September 15, 2008, https://www.washingtonpost.com/wp-dyn/content/article/2008/09/14/AR2008091401640.html.

209 Haidt, *The Righteous Mind*, 244.

210 Shepard, *The Tender Carnivore and the Sacred Game*, xvii.

211 Shepard, *The Tender Carnivore and the Sacred Game*, xviii.

212 Shepard, *The Tender Carnivore and the Sacred Game*, 4.

213 Shepard, *The Tender Carnivore and the Sacred Game*, 26.

214 Shepard, *The Tender Carnivore and the Sacred Game*, 35.

215 Shepard, *The Tender Carnivore and the Sacred Game*, 36.

216 Petersen, quoting a friend in *Heartsblood*, 71.

217 Nelson, "Finding Common Ground," 3.

218 Jonathan Parker, "The Camera or the Gun," in *Hunting—Philosophy for Everyone*, 169–70.

219 T. J. Kover, "Flesh, Death, and Tofu," in *Hunting—Philosophy for Everyone*, 178.

220 Barry Lopez, "From Arctic Dreams," in *A Hunter's Heart*, 315.

221 Brian Seitz, "Hunting for Meaning," in *Hunting—Philosophy for Everyone*, 69.

222 Petersen, *Heartsblood*, 177–78.

223 Paul Shepard, *Thinking Animals: Animals and the Development of Human Intelligence* (Athens, GA: University of Georgia Press, 1998), 13.

224 Shepard, *Thinking Animals*, 65.

225 Shepard, *Thinking Animals*, 71.

226 Dan Gardner, *The Science of Fear: How the Culture of Fear Manipulates Your Brain* (New York: Dutton, 2009), 174.

227 Petersen, *Heartsblood*, 128.

228 Petersen, *Heartsblood*, 47.

229 Martin, "Lionizing Cecil Makes Us Feel Good."

230 Martin, "Lionizing Cecil Makes Us Feel Good."

231 Nel Neddings, *Women and Evil* (Oakland, CA: University of California Press, 1991), 124.

232 Ruth Rudner, "The Call of the Climb," in *A Hunter's Heart*, 48.

233 "Mary Zeiss Stange," Center for Humans and Nature, accessed October 16, 2021, https://www.humansandnature.org/mary-zeiss-stange.

234 Stange, *Woman the Hunter*, 162.

235 Stange, *Woman the Hunter*, 129.

236 Stange, *Woman the Hunter*, 130.

237 Mitchell, "Why More Women Are Taking Up Hunting."

238 Kylie Vess, "Number of Women Shooters and Hunters on the Rise," *NRA Family* (website), March 2016, https://www.nrafamily.org/content/number-of-women-shooters-and-hunters-on-the-rise/.

239 "Infographic: Girl Power," *NSSF* (website), January 21, 2015, https://www.nssf.org/articles/infographic-girl-power/.

240 Mitchell, "Why More Women Are Taking Up Hunting."

241 Stephen Castle, "'Just See How I Shoot.' In Norway, Women Are Joining the Hunt," *The New York Times*, March 2017, https://www.nytimes.com/2017/03/28/world/europe/norway-female-hunters.html.

242 Stange, *Woman the Hunter*, 76.

243 Shepard, *Thinking Animals*, 242.

244 Shepard, *Thinking Animals*, 243.

245 Ted Kerasote, "Restoring the Older Knowledge," *Orion*, Winter 1996, https://www.kerasote.com/essays/RestoringOlderKnowledge.pdf.

246 Kerasote, "Restoring the Older Knowledge."

247 Kerasote, "Restoring the Older Knowledge."

248 "Ivan Carter Just Made Anti-Hunters an Awesome Offer," *Sporting Classic Daily*, August 7, 2015, https://sportingclassicsdaily.com/ivan-carter-just-made-anti-hunters-an-awesome-offer.

249 Alex Robinson, "Jim Shockey Goes Home: The Legacy of Modern Hunting's Most Influential Celebrity," *Outdoor Life*, November 20, 2017, https://www.outdoorlife.com/the-legacy-of-jim-shockey/.

250 David Smith, "Jim Shockey Talks with *Wide Open Spaces*—Part One," *Wide Open Spaces* (website), December 11, 2017, https://www.wideopenspaces.com/jim-shockey-talks-wide-open-spaces-part/.

251 Smith, "Jim Shockey Talks with *Wide Open Spaces*."

252 Mogomotsi Magome, "South Africa Seizes Rhino Horns to Be Smuggled to Malaysia," *AP News* (website), February 4, 2021, https://apnews.com/article/africa-south-africa-johannesburg-b577b0baaafb4e40b40f64eb7cdd20cd.

253 Bhaskar Vira, "Kaziranga's Ruthless Rangers Have Reduced Rhino Poaching by Simply Gunning Down Poachers at Sight," *Quarz India*, February 13, 2017, https://qz.com/india/908867/kazirangas-ruthless-rangers-have-reduced-rhino-poaching-by-simply-gunning-down-poachers-at-sight/.

254 "Key Facts about Hunters and Shooters: Did You Know?" NSSF, The Firearm Industry Trade Association, accessed October 18, 2021, https://www.letsgohunting.org/articles/key-facts-about-hunters-and-shooters/. For information on hunting and conservation, visit the National Hunting and Fishing Day website, https://nhfday.org/hunting/.

255 Frances Stead Sellers, "Hunting Is 'Slowly Dying Off,' and That Has Created a Crisis for the Nation's Many Endangered Species," *The Washington Post*, February 3, 2020, https://www.washingtonpost.com/national/hunting-is-slowly-dying-off-and-that-has-created-a-crisis-for-the-nations-public-lands/2020/02/02/554f51ac-331b-11ea-a053-dc6d944ba776_story.html.

256 Canadian Sporting Arms and Ammunition Association, "Canadian market overview," 2019, https://www.csaaa.org/wp-content/uploads/2020/01/CSAAA_Canadian-Market-Overview.pdf.

257 "Hunting in Europe Is Worth 16 Billion Euros," Hunters of Europe: FACE, September 29, 2016, https://www.face.eu/2016/09/hunting-in-europe-is-worth-16-billion-euros/.

258 William C. Gregory, "Latin American Consumptive Wildlife Tourism: An Analysis of the Industry as a Tool for Development," abstract (PhD diss., Baylor University Honors Program, 2014), https://baylor-ir.tdl.org/bitstream/handle/2104/9008/ThesisFinal.2%20pdf.pdf?sequence=1&isAllowed=y.

259 "Namibia: Facts," WWF, World Wildlife Fund, accessed October 18, 2021, https://www.worldwildlife.org/places/namibia.

260 Larry Keane, "Michigan Bald Eagle Comeback Is Win for Hunters," NSSF, National Firearms Industry Association, August 17, 2020, https://www.nssf.org/articles/michigan-bald-eagle-comeback-is-win-for-hunters/.

261 John Branch, "The Ultimate Pursuit in Hunting: Sheep," *The New York Times*, February 16, 2017, https://www.nytimes.com/2017/02/16/sports/bighorn-sheep-hunting.html.

262 *Angry Inuk*, directed by Alethea Arnaquq-Baril (2016; Montreal: EyeSteelFilm), 85 minutes.

263 Terry Tempest Williams, "An Open Letter to Major Wesley Powell," in *The Hour of Land: Personal Topography of America's National Parks* (New York, NY: Farrar, Straus and Giroux, 2016), 289.

264 Petersen, *A Hunter's Heart*, 157.

265 "Infographic: How Wildlife Is Thriving Because of Guns and Hunting," *NSSF* (website), March 26, 2014, https://www.nssf.org/articles/infographic-how-wildlife-is-thriving-because-of-guns-hunting/. Winter Park High, Science 122, "Environmental Science Case Studies," 77, https://www.coursehero.com/file/p628p2il/are-an-estimated-30-million-deer-in-the-United-States-today-Over-6-million-deer/.

266 Smith, "Jim Shockey Talks with *Wide Open Spaces*."

267 Branch, "The Ultimate Pursuit in Hunting."

268 Branch, "The Ultimate Pursuit in Hunting."

269 Curt Meine, "What's So New about the 'New Conservation'?" in *Keeping the Wild: Against the Domestication of Earth*, ed. G. Wuerthner, E. Crist, and T. Butler (Washington, DC: Island Press, 2014), 45–54.

270 "WWF Policy and Considerations on Trophy Hunting," July 2016, https://wwfint.awsassets.panda.org/downloads/wwf_policy_and_considerations_re_trophy_hunting__july_2016_.pdf.

271 Smith, "Jim Shockey Talks with *Wide Open Spaces*."

272 Eric Bowman, "Is Dopamine to Blame for Our Addictions?" *The Conversation*, December 3, 2015, https://theconversation.com/explainer-what-is-dopamine-and-is-it-to-blame-for-our-addictions-51268.

273 Bowman, "Is Dopamine to Blame for Our Addictions?"

274 Shepard, *Thinking Animals*, 38.

275 Shepard, *Thinking Animals*, 47.

276 Petersen, *Heartsblood*, 1.

277 Petersen, *Heartsblood*, 36.

278 Snyder, *The Practice of the Wild*, 65.

279 Snyder, *The Practice of the Wild*, 184.

280 Boer Deng, "A Time to Be Born: Why Do Birth Rates Peak at Different Times in Different Places?" *Slate*, April 28, 2014, https://slate.com/technology/2014/04/birth-rates-vary-by-season-and-latitude-what-explains-the-peaks.html.

281 Ivan Carter, "Proud to Be a Part of the Bubye Valley Conservancy Lion Project," Facebook, November 21, 2017, https://www.facebook.com/ivancartersafrica/photos/proud-to-be-a-part-of-the-bubye-valley-conservancy-lion-project-a-project-that-i/1534113733344927/.

282 Chris Mooney, "Just Looking at Nature Can Help Your Brain Work Better, Study Finds," *The Washington Post*, May 26, 2015, https://www.washingtonpost.com/news/energy-environment/wp/2015/05/26/viewing-nature-can-help-your-brain-work-better-study-finds/.

283 Mooney, "Just Looking at Nature Can Help Your Brain Work Better."

284 John Wendle, "Animals Rule Chernobyl Three Decades after Nuclear Disaster," *National Geographic* (website), April 18, 2016, https://www.nationalgeographic.com/animals/article/060418-chernobyl-wildlife-thirty-year-anniversary-science.

285 Dave Forman, "The Myth of the Humanized Pre-Columbian Landscape," in *Keeping the Wild*, 123.

286 Brendan Mackey, "The Future of Conservation," in *Keeping the Wild*, 132.

287 Meadowcreek, "Go to the Wilderness," Resilience Project Seminar (blog), December 17, 2017, https://meadowcreekvalley.wordpress.com/2017/12/17/wilderness/.

288 Richard K. Nelson, "Introduction: Finding Common Ground," in *A Hunter's Heart*, 2.

289 Petersen, *Heartsblood*, 15.

290 Shepard, *Thinking Animals*, 243.

291 Shepard, *Thinking Animals*, 261.

292 Petersen, *Heartsblood*, 42.

293 Petersen, *Heartsblood*, 235.

294 Petersen, *Heartsblood*, 235.

295 Stange, *Woman the Hunter*, 169.

296 Josh Honeycutt, "Devout Anti-Hunter Becomes Avid Bowhunter," Realtree (website), updated November 26, 2018, https://www.realtree.com/brow-tines-and-backstrap/devout-anti-hunter-becomes-avid-bowhunter.

297 Shepard, *Thinking Animals*, 212.

298 Dave Foreman, "The Myth of the Humanized Pre-Columbian Landscape," in *Keeping the Wild*, 116.

299 Shepard, *Thinking Animals*, 212.

300 Rick Bass, "An Appeal to Hunters," in *A Hunter's Heart*, 194.

301 Haidt, *The Righteous Mind*, 366.

302 "The Criteria for Selection," UNESCO World Heritage Centre, accessed June 23, 2022, https://whc.unesco.org/en/criteria/.

303 "Delta Waterfowl University Hunting Program," *Delta Waterfowl* (website), accessed May 24, 2022, https://deltawaterfowl.org/deltas-university-hunting-program/.

304 There are a few nicely gray areas on the Namibia-Kenya story. In an interview, a frequent visitor to modern Kenya told us this: There is an eight hundred fifty thousand-acre, now community-owned, conservancy in northern Kenya (Namunyak Wildlife Conservancy, https://sarara.co/). When hunting was banned with little warning in 1977, it directly led to a mass slaughter of the wild animals. This was by both the Indigenous (Samburu) and by Somalis. It took about ten years until virtually all the wildlife was removed. Fast forward to today—the lodges are community-owned, and international tourism supports the entire community. I have never seen so many giraffes in one place—teeming with wildlife. Elephants, leopards

and so much more are thriving. There are many ways "to skin the cat" of maintaining wild places. You have seen, with your own eyes, the Namibia model—but I have seen, with my own eyes, the Kenya model. The "fight" is not between current day Kenya and Namibia but between the protection/growth of the wild and the loss of the wild.

305 Petersen, *Heartsblood*, 70.

306 David Smith, "Jim Shockey Talks with *Wide Open Spaces*."

Suggested Reading

Asken, Michael J., Lt. Col. Dave Grossman, with Loren, W. Christensen. *Warrior Mindset*. Belleview, IL: Warrior Science Publications, 2010.

Becker, Ernest. *The Denial of Death*. New York, NY: Free Press Paperbacks (Simon & Schuster), 1997.

Cartmill, Matt. *A View to a Death in the Morning: Hunting and Nature through History*. Cambridge, MA: Harvard University Press, 1996.

Christensen, Loren W. *Mental Rehearsal for Warriors*. Portland, OR: LWC Books, 2014.

De Becker, Gavin. *The Gift of Fear: Survival Signals That Protect Us from Violence*. New York, NY: Random House, Inc., 1998.

Hunter, John A. *Hunter*. Long Beach, CA: Safari Press Inc., 1999.

———. *Hunter's Tracks*. Long Beach, CA: Safari Press Inc., 2014.

———. *White Hunter*. Long Beach, CA: Safari Press Inc., 1986.

Hunter, John A., and Daniel P. Mannix. *Tales of the African Frontier*. Long Beach, CA: Safari Press Inc., 2012.

Kerasote, Ted. *Bloodties: Nature, Culture, and the Hunt.* New York, NY: Random House, 2013. Kindle Edition.

Nelson, Richard K. *Make Prayers to the Raven: A Koyukon View of the Northern Forest.* Chicago, IL: The University of Chicago Press, 1986.

Roosevelt, Theodore. *African Game Trails: An Account of the African Wanderings of an American Hunter-naturalist.* New York, NY: St. Martin's Press, 1988.

Shockey, Jim. *Ultimate Big Game Adventures: Wild Hunts across North America.* Victoria, BC: Folkart Interiors, 2003.

———. *Ultimate Big Game Adventures—Part II: More Wild Hunts across North America.* Victoria, BC: Folkart Interiors, 2003.

Selected Bibliography

Books

Bird, Chris. *Thank God I Had a Gun: True Accounts of Self-Defense*. San Antonio, TX: Privateer Publications, 2007.

Boyden, Joseph. *Three Day Road*. Toronto, ON: Penguin Group, 2005.

Brody, Hugh. *The Other Side of Eden: Hunters, Farmers, and the Shaping of the World*. New York, NY: North Point Press, a division of Farrar, Straus and Giroux, 2001.

Capstick, Peter Hathaway. *Death in the Long Grass*. New York, NY: St. Martin's Press, 1977.

———. *Death in the Silent Places*. New York, NY: St. Martin's Press, 1981.

Cooper, Jeff. *Principles of Personal Defense*. Boulder, CO: Paladin Press, 2006.

Diamond, Jared. *The Third Chimpanzee: The Evolution and Future of the Human Animal*. New York, NY: HarperCollins Publishers, 1992 (Harper Perennial edition 1993).

———. *Why Is Sex Fun?: The Evolution of Human Sexuality*. New York, NY: Basic Books (Perseus Books Group), 1997.

Ehrenreich, Barbara. *Blood Rites: Origins and History of the Passions of War*. New York, NY: Henry Holt and Company, 1997.

Gardner, Dan. *The Science of Fear: How the Culture of Fear Manipulates Your Brain*. New York, NY: Plume (Penguin Group), 2008.

Grossman, Lt. Col. Dave. *On Killing: The Psychological Cost of Learning to Kill in War and Society*. New York, NY: Time Warner Book Group, Back Bay Books (Little, Brown and Company), 1995.

Grossman, Lt. Col. Dave with Loren W. Christensen. *On Combat: The Psychology and Physiology of Deadly Conflict in War and in Peace*. Milstadt, IL: Warrior Science Publications, 2008.

Haidt, Jonathan. *The Righteous Mind: Why Good People Are Divided by Politics and Religion*. New York, NY: Pantheon Books (Random House, Inc.), 2012.

Harari, Yuval Noah. *Sapiens: A Brief History of Humankind*. Toronto, ON: Signal Books, 2014.

Holthouse, Hector. *River of Gold: The Wild Days of the Palmer River Gold Rush*. Sydney, Australia: HarperCollins, 1967.

Kowalsky, Nathan, ed. *Hunting—Philosophy for Everyone: In Search of the Wild Life*. Chichester, West Sussex, UK: Blackwell Publishing Ltd. (John Wiley & Sons), 2010.

Lee, Richard B., and Richard H. Daly, eds. *The Cambridge Encyclopedia of Hunters and Gatherers*. Cambridge, UK: Cambridge University Press, 1999.

Leopold, Aldo. *A Sand County Almanac: And Sketches Here and There*. New York, NY: Ballantine Books, 1966.

Levi-Strauss, Claude. *Myth and Meaning*. New York, NY: Schocken Books, 1979.

Miller, Linda K., and Keith A. Cunningham. *Secrets of Mental Marksmanship: How to Fire Perfect Shots*. Mascoutah, IL: Killology Research Group, 2018.

Milke, Mark. *The Victim Cult: How the Culture of Blame Hurts Everyone and Wrecks Civilizations*. Victoria, BC: Thomas and Black, 2019.

Neddings, Nel. *Women and Evil*. Oakland, CA: University of California Press, 1991.

Ortega y Gasset, José. *Meditations on Hunting*. Belgrade, MT: Wilderness Adventures Press, Inc., 1995.

Petersen, David. *Heartsblood: Hunting, Spirituality, and Wildness in America*. Durango, CO: Raven's Eye Press, 2010.

Petersen, David, ed. *A Hunter's Heart: Honest Essays on Blood Sport*. New York, NY: Henry Holt and Company, 1997.

Peterson, Jordan B. *12 Rules for Life: An Antidote to Chaos*. Toronto, ON: Random House Canada, 2018.

Roosevelt, Theodore. *Outdoor Pastimes of an American Hunter*. New York, NY: Charles Scribner's, 1905.

Shepard, Paul. *The Tender Carnivore and the Sacred Game*. Athens, GA: University of Georgia Press, 1998.

———. *Thinking Animals: Animals and the Development of Human Intelligence*. Athens, GA: University of Georgia Press, 1998.

Siddle, Bruce K. *Sharpening the Warrior's Edge*. Belleville, IL: PPCT Research Publications, 1995.

Snyder, Gary. *The Practice of the Wild: Essays*. Berkley, CA: Counterpoint Press, 1990.

Stange, Mary Zeiss. *Woman the Hunter*. Boston, MA: Beacon Press, 1997.

Thoreau, Henry David. *Walden, or Life in the Woods*. New York, NY: Signet Classics, 2004.

Van Horne, Patrick, Jason Riley. *Left of Bang: How the Marine Corps' Combat Hunter Program Can Save Your Life*. New York, NY: Black Irish Entertainment LLC, 2014.

Wuerthner, George, Eileen Crist, and Tom Butler. *Keeping the Wild: Against the Domestication of Earth*. San Francisco, CA: Foundation for Deep Ecology, 2014.

Print Articles

Dreher, Jean-Claude, Simon Dunne, Agnieszka Pazderska, et al. "Testosterone Causes Both Prosocial and Antisocial Status-enhancing Behaviors in Human Males." *Proceedings of the National Academy of Sciences of the United States of America* 113, no. 41 (2016): 11633–38. doi:10.1073/pnas.1608085113.

Kellert, Stephen. "Attitudes and Characteristics of Hunters and Antihunters." *International Journal for the Study of Animal Problems* (1980): 87–119.

Löw, Andreas, Peter J. Lang, J. Carson Smith, and Margaret M. Bradley. "Both Predator and Prey: Emotional Arousal in Threat and Reward." *Psychological Science*, 19.9 (September 1, 2008): 865–73.

Magome, Mogomotsi. "South Africa Seizes Rhino Horns to Be Smuggled to Malaysia." *AP News* (website). February 4, 2021. https://apnews.com/article/africa-south-africa-johannesburg-b577b0baaafb4e40b40f64eb7cdd20cd.

Miller, Linda, and Keith Cunningham. "Coming Home to Namibia." *Ontario Out of Doors* 48, no. 1 (Jan.–Feb. 2016): 42–43.

Stanford, Craig B. "Chimpanzee Hunting Behavior and Human Evolution." *American Scientist* 83.3 (1995): 256–61.

Online Resources

Arnaquq-Baril, Alethea, director. *Angry Inuk*. 2016; Montreal: EyeSteelFilm. 85 minutes.

Branch, John. "The Ultimate Pursuit in Hunting: Sheep." *The New York Times*. February 16, 2017. https://www.nytimes.com/2017/02/16/sports/bighorn-sheep-hunting.html.

Bryner, Jeanna. "Modern Humans Retain Caveman's Survival Instincts." *Live Science*. September 24, 2007. https://www.livescience.com/4631-modern-humans-retain-caveman-survival-instincts.html.

"Delta Waterfowl University Hunting Program." *Delta Waterfowl* (website). Accessed May 24, 2022. https://deltawaterfowl.org/deltas-university-hunting-program/.

Geist, Valerius. "Beyond Wolf Advocacy, toward Realistic Policies for Carnivore Conservation." Boon and Crockett Club (website). Winter 2008. https://www.boone-crockett.org/beyond-wolf-advocacy-toward-realistic-policies-carnivore-conservation.

Honeycutt, Josh. "Devout Anti-Hunter Becomes Avid Bowhunter." Realtree (website). Updated November 26, 2018. https://www.realtree.com/brow-tines-and-backstrap/devout-anti-hunter-becomes-avid-bowhunter.

"Hunting." The Jane Goodall Institute UK. Accessed October 16, 2021. https://www.janegoodall.org.uk/chimpanzees/chimpanzee-central/15-chimpanzees/chimpanzee-central/20-hunting.

"Hunting in Europe Is Worth 16 Billion Euros." Hunters of Europe: FACE. September 29, 2016. https://www.face.eu/2016/09/hunting-in-europe-is-worth-16-billion-euros/.

"Ivan Carter Just Made Anti-Hunters an Awesome Offer." *Sporting Classic Daily*. August 7, 2015. https://sportingclassicsdaily.com/ivan-carter-just-made-anti-hunters-an-awesome-offer.

Kerasote, Ted. "Restoring the Older Knowledge." *Orion*. Winter 1996. https://www.kerasote.com/essays/RestoringOlderKnowledge.pdf.

Kipling, Rudyard. "The Law of the Jungle." Accessed May 24, 2022. http://www.kiplingsociety.co.uk/poems_lawofjungle.htm.

Leonard, William R., and Marcia L. Robertson. "Evolutionary Perspectives on Human Nutrition: The Influence of Brain and Body Size on Diet and Metabolism." *American Journal of Human Biology* Volume 6.1 (1994): 77–78. https://onlinelibrary.wiley.com/doi/10.1002/ajhb.1310060111.

Martin, Glen. "Kenya: Wildlife Loved to Death." *African Indaba* Volume 12.2 (March 2014). http://www.africanindaba.com/2014/03/kenya-wildlife-loved-to-death-march-2014-volume-12-2/.

———. "Lionizing Cecil Makes Us Feel Good, but a Trophy Hunting Ban Will Accelerate Slaughter." *California* (website). CalAlumni Association at U.C. Berkeley. August 3, 2015. https://alumni.berkeley.edu/california-magazine/justin/2021-08-25/lionizing-cecil-makes-us-feel-good-trophy-hunting-ban-will.

McKie, Robin. "Humans Hunted for Meat 2 Million Years Ago." *The Guardian*. September 22, 2012. https://www.theguardian.com/science/2012/sep/23/human-hunting-evolution-2million-years.

Robinson, Alex. "Jim Shockey Goes Home: The Legacy of Modern Hunting's Most Influential Celebrity." *Outdoor Life*. Nov. 20, 2017. https://www.outdoorlife.com/the-legacy-of-jim-shockey/.

Sellers, Frances Stead. "Hunting Is 'Slowly Dying Off,' and That Has Created a Crisis for the Nation's Many Endangered Species." *The Washington Post*. February 3, 2020. https://www.washingtonpost.com/national/hunting-is-slowly-dying-off-and-that-has-created-a-crisis-for-the-nations-public-lands/2020/02/02/554f51ac-331b-11ea-a053-dc6d944ba776_story.html.

Smith, David. "Jim Shockey Talks with *Wide Open Spaces*—Part One." *Wide Open Spaces* (website). December 11, 2017. https://www.wideopenspaces.com/jim-shockey-talks-wide-open-spaces-part/.

Stanford, Craig B. "The Predatory Behavior and Ecology of Wild Chimpanzees." Archived at Gambassa. Accessed October 6, 2021. http://www.gambassa.com/public/Collaborations/1264/850/4911/Stanford,2011.html.

Taub, Benjamin. "Why Are Herbivorous Prairie Dogs So Murderous?" IFL Science. March 24, 2016. https://www.iflscience.com/plants-and-animals/why-do-some-herbivores-kill-other-animals/.

"Theoretical Tiger Chases Statistical Sheep to Probe Immune System Behavior: Physicists Update Predator-Prey Model for More Clues on How Bacteria Evade Attack from Killer Cells." ScienceDaily. April 29, 2016. https://www.sciencedaily.com/releases/2016/04/160429095152.htm.

"Wolf Attack." Wikipedia. Last modified October 4, 2021. https://en.wikipedia.org/wiki/Wolf_attack.

Index

392 → On Hunting

394 → On Hunting

denial of, 76, 97–98, 270, 318

humane, 156–57, 187, 209

hunter's connection to, 16, 68–69, 97–98, 109–10, 266, 346–47, 348

hunter's respect for, 270, 277, 314, 345, 348

life-and-death situations, 195–96, 230–31, 233

deep ecology, 44, 300

Delta's University Hunting Program, 340

Disney, 82, 145, 252, 264, 275, 305, 351

domesticated animals, 88, 95, 99, 100, 181–82, 265, 302

dopamine, 38–39, 243, 314–16

ecology, 99, 191, 239, 248, 259, 276, 300, 309–10, 315, 330

economics, 288–90, 309, 330

About the Authors

Lt. Col. Dave Grossman

Dave is an award-winning author and nationally recognized as a powerful, dynamic speaker. He has authored over a dozen books, including his perennial bestsellers *On Killing*, *On Combat*, and *On Spiritual Combat*, a *New York Times* best-selling book coauthored with Glenn Beck, and many other successful books and scholarly papers. His books are required or recommended reading in all four branches of the US Armed Forces and in federal and local law enforcement academies nationwide.

He is a US Army Ranger, a paratrooper, a prior service sergeant, and a former West Point psychology professor. He has five patents to his name, has earned a black belt in Hojutsu (the martial art of the firearm), and has been inducted into the USA Martial Arts Hall of Fame.

Dave's research was cited by the President of the United States in a national address, he has testified before the US Senate, the US Congress, and numerous state legislatures, and he has been invited to the White House on two occasions to brief the president and the vice president in his areas of expertise.

Since his retirement from the US Army in 1998, Dave has been on the road over two hundred days a year for over twenty-four years as one of the world's leading trainers for military, law enforcement, mental health providers, and school safety organizations. He has been inducted as a Life Diplomate by the American Board for Certification in Homeland Security and a Life Member of the American College of Forensic Examiners Institute.

Linda K. Miller

Linda has over twenty-five years of business experience, including management consulting, business planning, business management, marketing, and information systems.

She has considerable experience in international small-bore target shooting as a member of Canada's Shooting Team. She has won medals in the Commonwealth Games 1994, Cuba World Cup 1995, and Mexico World Cup 1993. In 1999, Linda became the first woman to win the Ontario Lieutenant Governor's Medal for shooting (full-bore rifle); these competitions have a proud and honored history of over 140 years. Linda is also the first and only woman to be the National Sniper Rifle Champion (2008). She holds many provincial and national titles and records, and she has been a member of several Canadian teams to international championships throughout the world.

Linda is an accomplished and internationally certified shooting coach. She has coached provincial and national teams (small-bore and full-bore) as well as the Canadian CISM (Conseil International du Sport Militaire) team. With Keith, she has coached the Canadian Forces Combat Shooting Team to many honors in England and Australia. They have coached thirteen members of the military to a Queen's Medal, the top award for marksmanship within the Canadian Forces. Linda has also volunteered as a director, manager, administrator, and consultant in Club, Provincial, and National shooting sports organizations.

Linda started hunting in 1995 and has since hunted small and big game in Canada (Ontario), the USA (Georgia, South Dakota), and Africa (Namibia).

Capt. Keith A. Cunningham

Keith is a career military officer with a combined experience of over twenty-five years with the Canadian Armed Forces and the US Army. He has considerable practical experience, including a combat tour in Vietnam, where he was a US Army Ranger specializing as a sniper and in long-range reconnaissance. In the Canadian Army, he was a part of the Special Service Force. He did peacekeeping and counter-sniper operations in Cyprus and annual unit and command-level military exercises in North America and Europe.

Keith has taught marksmanship courses at the Canadian Forces Infantry School and at many police forces in Ontario. He was a certified instructor/examiner for the Firearms Safety Education Service of Ontario, and he was a Hunter Safety

Instructor/Examiner. Keith is an internationally certified shooting coach and has successfully coached teams to national and international excellence. He is currently the Chief Instructor with the MilCun Training Center.

Keith is a qualified gunsmith with over thirty years of experience, specializing in long-range practical rifles. He has built and regulated rifles used by Canadian police agencies and international competitors around the world. He is an internationally renowned rifle and pistol competitor, having won honors at Bisley, the World Long Range Championships, and the Commonwealth Games. He has been the Canadian National Champion in pistol, tactical/service rifle, sniper rifle, and 3-Gun a total of nineteen times. He is a member of the Canadian Forces Sports Hall of Fame and a three-time member of the Dominion of Canada Rifle Association Hall of Fame (in the target rifle, service rifle, and builder categories).

Keith has hunted since he was a youngster growing up in rural Canada. He has since hunted small and big game in Canada (Ontario), the USA (Georgia, South Dakota), and Africa (Namibia).

Keith and Linda together built the MilCun Training Center, where they run many professional and recreational shooting courses and events. They are popular guest lecturers and speakers, providing seminars and courses to police, military, and civilian marksmen in Canada and internationally. Their articles on marksmanship have been published in shooting magazines such as *Precision Shooting, Tactical Shooter, The Accurate Rifle, Canadian Marksman*, the *Canadian Forces Infantry Journal, Ontario Out of Doors*, and *AIM Magazine*. They are the authors of *The Wind Book for Rifle Shooters* and *Secrets of Mental Marksmanship*. They are currently working on several books.